Solutions Manual
to Accompany
Nonlinear Programming:
Theory and Algorithms

Solutions Manual to Accompany Nonlinear Programming: Theory and Algorithms

Third Edition

Mokhtar S. Bazaraa
Department of Industrial and Systems Engineering
Georgia Institute of Technology
Atlanta, GA

Hanif D. Sherali
Department of Industrial and Systems Engineering
Virginia Polytechnic Institute and State University
Blacksburg, VA

C. M. Shetty
Department of Industrial and Systems Engineering
Georgia Institute of Technology
Atlanta, GA

Solutions Manual Prepared by:

Hanif D. Sherali

Joanna M. Leleno

Acknowledgment: This work has been partially supported by the National Science Foundation under Grant No. CMMI-0969169.

For general information on our other products and services please contact our Customer Care
Department within the United States at (800) 762-2974, outside the United States at (317) 572-3993
or fax (317) 572-4002.

Wiley also publishes its books in a variety of electronic formats. Some content that appears in print,
however, may not be available in electronic formats. For more information about Wiley products,
visit our web site at www.wiley.com.

Library of Congress Cataloging-in-Publication Data is available.

ISBN 978-1-118-76237-0

10 9 8 7 6 5 4 3 2 1

TABLE OF CONTENTS

Chapter 1: Introduction 1

    1.1, 1.2, 1.4, 1.6, 1.10, 1.13

Chapter 2 Convex Sets... 4

    2.1, 2.2, 2.3, 2.7, 2.8, 2.12, 2.15, 2.21, 2.24, 2.31, 2.42, 2.45,
    2.47, 2.49, 2.50, 2.51, 2.52, 2.53, 2.57

Chapter 3: Convex Functions and Generalizations 15

    3.1, 3.2, 3.3, 3.4, 3.9, 3,10, 3.11, 3.16, 3.18, 3.21, 3.22, 3.26,
    3.27, 3.28, 3.31, 3.37, 3.39, 3.40, 3.41, 3.45, 3.48, 3.51, 3.54,
    3.56, 3.61, 3.62, 3.63, 3.64, 3.65

Chapter 4: The Fritz John and Karush-Kuhn-Tucker Optimality Conditions .. 29

    4.1, 4.4, 4.5, 4.6, 4.7, 4.8, 4.9, 4.10, 4.12, 4.15, 4.27, 4.28, 4.30,
    4.31, 4.33, 4.37, 4.41, 4.43

Chapter 5: Constraint Qualifications... 46

    5.1, 5.12, 5.13, 5.15, 5.20

Chapter 6: Lagrangian Duality and Saddle Point Optimality Conditions 51

    6.2, 6.3, 6.4, 6.5, 6.7, 6.8, 6.9, 6.14, 6.15, 6.21, 6.23, 6.27, 6.29,

Chapter 7: The Concept of an Algorithm... 64

    7.1, 7.2, 7.3, 7.6, 7.7, 7.19

Chapter 8: Unconstrained Optimization... 69

    8.10, 8.11, 8.12, 8.18, 8.19, 8.21, 8.23, 8.27, 8.28, 8.32, 8.35,
    8.41, 8.47, 8.51, 8.52

Chapter 9: Penalty and Barrier Functions 88

    9.2, 9.7, 9.8, 9.12, 9.13, 9.14, 9.16, 9.19, 9.32

Chapter 10: Methods of Feasible Directions....................................... 107

    10.3, 10.4, 10.9, 1.012, 10.19, 10.20, 10.25, 10.33, 10.36, 10.41,
    10.44, 10.47, 10.52

Chapter 11: Linear Complementary Problem, and Quadratic, Separable,
Fractional, and Geometric Programing 134

11.1, 11.5, 11.12, 11.18, 11.19, 11.22, 11.23, 11.24, 11.36, 11.41,
11.42, 11.47, 11.48, 11.50, 11.51, 11.52

CHAPTER 1:

INTRODUCTION

1.1 In the figure below, x_{min} and x_{max} denote optimal solutions for Part (a) and Part (b), respectively.

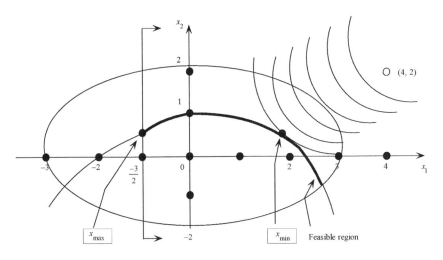

1.2 a. The total cost per time unit (day) is to be minimized given the storage limitations, which yields the following model:

$$\text{Minimize} \quad f(Q_1, Q_2) = k_1 \frac{d_1}{Q_1} + h_1 \frac{Q_1}{2} + k_2 \frac{d_2}{Q_2} + h_2 \frac{Q_2}{2} + c_1 d_1 + c_2 d_2$$

$$\text{subject to} \quad s_1 Q_1 + s_2 Q_2 \le S$$

$$Q_1 > 0, \ Q_2 > 0.$$

Note that the last two terms in the objective function are constant and thus can be ignored while solving this problem.

b. Let S_j denote the lost sales (in each cycle) of product j, $j = 1, 2$. In this case, we replace the objective function in Part (a) with $F(Q_1, Q_2, S_1, S_2)$, where $F(Q_1, Q_2, S_1, S_2) = F_1(Q_1, S_1) + F_2(Q_2, S_2)$, and where

$$F_j(Q_j, S_j) = \frac{d_j}{Q_j + S_j}(k_j + c_j Q_j + \ell_j S_j - PQ_j) + h_j \frac{Q_j^2}{2(Q_j + S_j)}, \quad j = 1, 2.$$

This follows since the cycle time is $\dfrac{Q_j + S_j}{d_j}$, and so over some T

days, the number of cycles is $\dfrac{T d_j}{Q_j + S_j}$. Moreover, for each cycle, the

fixed setup cost is k_j, the variable production cost is $c_j Q_j$, the lost

sales cost is $\ell_j S_j$, the profit (negative cost) is $P Q_j$, and the

inventory carrying cost is $\dfrac{h_j}{2} Q_j (\dfrac{Q_j}{d_j})$. This yields the above total cost

function on a daily basis.

1.4 Notation: $\quad x_j$: production in period $j, j = 1,\ldots,n$

$\qquad\qquad\quad d_j$: demand in period $j, j = 1,\ldots,n$

$\qquad\qquad\quad I_j$: inventory at the end of period $j, j = 0, 1,\ldots,n$.

The production scheduling problem is to:

Minimize $\sum\limits_{j=1}^{n} [f(x_j) + c I_{j-1}]$

subject to

$\qquad x_j - d_j + I_{j-1} = I_j \quad$ for $j = 1,\ldots,n$

$\qquad I_j \le K \quad$ for $j = 1,\ldots,n-1$

$\qquad I_n = 0$

$\qquad x_j \ge 0, \ I_j \ge 0 \quad$ for $j = 1,\ldots,n-1$.

1.6 Let X denote the set of feasible portfolios. The task is to find an $x^* \in X$ such that there does not exist an $\bar{x} \in X$ for which $\bar{c}'\bar{x} \ge \bar{c}'x^*$ and $\bar{x}' V \bar{x} \le x^{*'} V x^*$, with at least one inequality strict. One way to find efficient portfolios is to solve:

Maximize $\{\mu_1 \bar{c}' x - \mu_2 x' V x : x \in X\}$

for different values of $(\mu_1, \mu_2) > 0$ such that $\mu_1 + \mu_2 = 1$.

1.10 Let x and p denote the demand and production levels, respectively, and let Z denote a standard normal random variable. Then we need p to be such that $P(p < x - 5) \le 0.01$, which by the continuity of the normal random variable is equivalent to $P(x \ge p + 5) \le 0.01$. Therefore, p must satisfy

2

$$P(Z \geq \frac{p + 5 - 150}{7}) \leq 0.01,$$

where Z is a standard normal random variable. From tables of the standard normal distribution we have $P(Z \geq 2.3267) = 0.01$. Thus, we want

$\frac{p - 145}{7} \geq 2.3267$, or that the chance constraint is equivalent to

$p \geq 161.2869$.

1.13 We need to find a positive number K that minimizes the expected total cost. The expected total cost is $\alpha(1 - p)P(\bar{x} \leq K \,|\, \mu = \mu_2) + \beta p P(\bar{x} > K \,|\, \mu = \mu_1)$. Therefore, the mathematical programming problem can be formulated as follows:

$$\text{Minimize} \quad \alpha(1 - p)\int_0^K f(\bar{x} \,|\, \mu_2)d\bar{x} + \beta p \int_0^\infty f(\bar{x} \,|\, \mu_1)d\bar{x}$$

subject to $K \geq 0$.

If the conditional distribution functions $F(\bar{x} \,|\, \mu_2)$ and $F(\bar{x} \,|\, \mu_1)$ are known, then the objective function is simply $\alpha(1 - p)F(K \,|\, \mu_2) + \beta p(1 - F(K \,|\, \mu_1))$.

CHAPTER 2:

CONVEX SETS

2.1 Let $x \in conv(S_1 \cap S_2)$. Then there exists $\lambda \in [0,1]$ and x_1, $x_2 \in S_1 \cap S_2$ such that $x = \lambda x_1 + (1 - \lambda)x_2$. Since x_1 and x_2 are both in S_1, x must be in $conv(S_1)$. Similarly, x must be in $conv(S_2)$. Therefore, $x \in conv(S_1) \cap conv(S_2)$. (Alternatively, since $S_1 \subseteq conv(S_1)$ and $S_2 \subseteq conv(S_2)$, we have $S_1 \cap S_2 \subseteq conv(S_1) \cap conv(S_2)$ or that $conv[S_1 \cap S_2] \subseteq conv(S_1) \cap conv(S_2)$.)

An example in which $conv(S_1 \cap S_2) \neq conv(S_1) \cap conv(S_2)$ is given below:

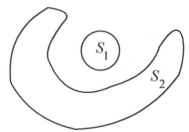

Here, $conv(S_1 \cap S_2) = \varnothing$, while $conv(S_1) \cap conv(S_2) = S_1$ in this case.

2.2 Let S be of the form $S = \{x : Ax \leq b\}$ in general, where the constraints might include bound restrictions. Since S is a polytope, it is bounded by definition. To show that it is convex, let y and z be any points in S, and let $x = \lambda y + (1 - \lambda)z$, for $0 \leq \lambda \leq 1$. Then we have $Ay \leq b$ and $Az \leq b$, which implies that

$$Ax = \lambda Ay + (1 - \lambda)Az \leq \lambda b + (1 - \lambda)b = b,$$

or that $x \in S$. Hence, S is convex.

Finally, to show that S is closed, consider any sequence $\{x_n\} \to x$ such that $x_n \in S$, $\forall n$. Then we have $Ax_n \leq b$, $\forall n$, or by taking limits as $n \to \infty$, we get $Ax \leq b$, i.e., $x \in S$ as well. Thus S is closed.

2.3 Consider the closed set S shown below along with $conv(S)$, where $conv(S)$ is not closed:

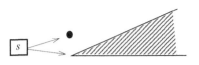 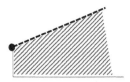

Now, suppose that $S \subseteq \mathbb{R}^p$ is closed. Toward this end, consider any sequence $\{x_n\} \to x$, where $x_n \in conv(S)$, $\forall n$. We must show that $x \in conv(S)$. Since $x_n \in conv(S)$, by definition (using Theorem 2.1.6), we have that we can write $x_n = \sum\limits_{r=1}^{p+1} \lambda_{nr} x_n^r$, where $x_n^r \in S$ for $r = 1,...,p+1$, $\forall n$, and where $\sum\limits_{r=1}^{p+1} \lambda_{nr} = 1$, $\forall n$, with $\lambda_{nr} \geq 0$, $\forall r,n$.

Since the λ_{nr}-values as well as the x_n^r-points belong to compact sets, there exists a subsequence K such that $\{\lambda_{nr}\}_K \to \lambda_r$, $\forall r = 1,...,p+1$, and $\{x_n^r\} \to x^r$, $\forall r = 1,...,p+1$. From above, we have taking limits as $n \to \infty$, $n \in K$, that

$$x = \sum_{r=1}^{p+1} \lambda_r x^r, \text{ with } \sum_{r=1}^{p+1} \lambda_r = 1, \ \lambda_r \geq 0, \ \forall r = 1,...,p+1,$$

where $x^r \in S$, $\forall r = 1,...,p+1$ since S is closed. Thus by definition, $x \in conv(S)$ and so $conv(S)$ is closed. $\square$

2.7 **a.** Let y^1 and y^2 belong to AS. Thus, $y^1 = Ax^1$ for some $x^1 \in S$ and $y^2 = Ax^2$ for some $x^2 \in S$. Consider $y = \lambda y^1 + (1-\lambda)y^2$, for any $0 \leq \lambda \leq 1$. Then $y = A[\lambda x^1 + (1-\lambda)x^2]$. Thus, letting $x \equiv \lambda x^1 + (1-\lambda)x^2$, we have that $x \in S$ since S is convex and that $y = Ax$. Thus $y \in AS$, and so, AS is convex.

b. If $\alpha \equiv 0$, then $\alpha S \equiv \{0\}$, which is a convex set. Hence, suppose that $\alpha \neq 0$. Let αx^1 and $\alpha x^2 \in \alpha S$, where $x^1 \in S$ and $x^2 \in S$. Consider $\alpha x = \lambda \alpha x^1 + (1-\lambda)\alpha x^2$ for any $0 \leq \lambda \leq 1$. Then, $\alpha x = \alpha[\lambda x^1 + (1-\lambda)x^2]$. Since $\alpha \neq 0$, we have that $x = \lambda x^1 + (1-\lambda)x^2$, or that $x \in S$ since S is convex. Hence $\alpha x \in \alpha S$ for any $0 \leq \lambda \leq 1$, and thus αS is a convex set.

2.8 $S_1 + S_2 = \{(x_1, x_2) : 0 \leq x_1 \leq 1, \ 2 \leq x_2 \leq 3\}$.

$$S_1 - S_2 = \{(x_1, x_2) : -1 \le x_1 \le 0, \ -2 \le x_2 \le -1\}.$$

2.12 Let $S = S_1 + S_2$. Consider any y, $z \in S$, and any $\lambda \in (0,1)$ such that $y = y_1 + y_2$ and $z = z_1 + z_2$, with $\{y_1, z_1\} \subseteq S_1$ and $\{y_2, z_2\} \subseteq S_2$. Then $\lambda y + (1 - \lambda)z = \lambda y_1 + \lambda y_2 + (1 - \lambda)z_1 + (1 - \lambda)z_2$. Since both sets S_1 and S_2 are convex, we have $\lambda y_i + (1 - \lambda)z_i \in S_i$, $i = 1, 2$. Therefore, $\lambda y + (1 - \lambda)z$ is still a sum of a vector from S_1 and a vector from S_2, and so it is in S. Thus S is a convex set.

Consider the following example, where S_1 and S_2 are closed, and convex.

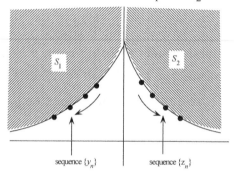

Let $x_n = y_n + z_n$, for the sequences $\{y_n\}$ and $\{z_n\}$ shown in the figure, where $\{y_n\} \subseteq S_1$, and $\{z_n\} \subseteq S_2$. Then $\{x_n\} \to 0$ where $x_n \in S$, $\forall n$, but $0 \notin S$. Thus S is not closed.

Next, we show that if S_1 is compact and S_2 is closed, then S is closed. Consider a convergent sequence $\{x_n\}$ of points from S, and let x denote its limit. By definition, $x_n = y_n + z_n$, where for each n, $y_n \in S_1$ and $z_n \in S_2$. Since $\{y_n\}$ is a sequence of points from a compact set, it must be bounded, and hence it has a convergent subsequence. For notational simplicity and without loss of generality, assume that the sequence $\{y_n\}$ itself is convergent, and let y denote its limit. Hence, $y \in S_1$. This result taken together with the convergence of the sequence $\{x_n\}$ implies that $\{z_n\}$ is convergent to z, say. The limit, z, of $\{z_n\}$ must be in S_2, since S_2 is a closed set. Thus, $x = y + z$, where $y \in S_1$ and $z \in S_2$, and therefore, $x \in S$. This completes the proof. $\square$

2.15 a. First, we show that $conv(S) \subseteq \hat{S}$. For this purpose, let us begin by showing that S_1 and S_2 both belong to $\hat{S}$. Consider the case of S_1 (the case of S_2 is similar). If $x \in S_1$, then $A_1x \le b_1$, and so, $x \in \hat{S}$ with $y = x$, $z = 0$, $\lambda_1 = 1$, and $\lambda_2 = 0$. Thus $S_1 \cup S_2 \subseteq \hat{S}$, and since $\hat{S}$ is convex, we have that $conv[S_1 \cup S_2] \subseteq \hat{S}$.

Next, we show that $\hat{S} \subseteq conv(S)$. Let $x \in \hat{S}$. Then, there exist vectors y and z such that $x = y + z$, and $A_1y \le b_1\lambda_1$, $A_2z \le b_2\lambda_2$ for some $(\lambda_1, \lambda_2) \ge 0$ such that $\lambda_1 + \lambda_2 = 1$. If $\lambda_1 = 0$ or $\lambda_2 = 0$, then we readily obtain $y = 0$ or $z = 0$, respectively (by the boundedness of S_1 and S_2), with $x = z \in S_2$ or $x = y \in S_1$, respectively, which yields $x \in S$, and so $x \in conv(S)$. If $\lambda_1 > 0$ and $\lambda_2 > 0$, then

$x = \lambda_1 y_1 + \lambda_2 z_2$, where $y_1 = \dfrac{1}{\lambda_1}y$ and $z_2 = \dfrac{1}{\lambda_2}z$. It can be easily verified in this case that $y_1 \in S_1$ and $z_2 \in S_2$, which implies that both vectors y_1 and z_2 are in S. Therefore, x is a convex combination of points in S, and so $x \in conv(S)$. This completes the proof ☐

b. Now, suppose that S_1 and S_2 are not necessarily bounded. As above, it follows that $conv(S) \subseteq \hat{S}$, and since $\hat{S}$ is closed, we have that $clconv(S) \subseteq \hat{S}$. To complete the proof, we need to show that $\hat{S} \subseteq clconv(S)$. Let $x \in \hat{S}$, where $x = y + z$ with $A_1y \le b_1\lambda_1$, $A_2z \le b_2\lambda_2$, for some $(\lambda_1, \lambda_2) \ge 0$ such that $\lambda_1 + \lambda_2 = 1$. If $(\lambda_1, \lambda_2) > 0$, then as above we have that $x \in conv(S)$, so that $x \in clconv(S)$. Thus suppose that $\lambda_1 = 0$ so that $\lambda_2 = 1$ (the case of $\lambda_1 = 1$ and $\lambda_2 = 0$ is similar). Hence, we have $A_1y \le 0$ and $A_2z \le b_2$, which implies that y is a recession direction of S_1 and $z \in S_2$ (if S_1 is bounded, then $y \equiv 0$ and then $x = z \in S_2$ yields $x \in clconv(S)$). Let $\bar{y} \in S_1$ and consider the sequence

$$x_n = \lambda_n[\bar{y} + \frac{1}{\lambda_n}y] + (1 - \lambda_n)z, \text{ where } 0 < \lambda_n \le 1 \text{ for all } n.$$

7

Note that $\bar{y} + \dfrac{1}{\lambda_n} y \in S_1$, $z \in S_2$, and so $x_n \in conv(S)$, $\forall n$.

Moreover, letting $\{\lambda_n\} \to 0^+$, we get that $\{x_n\} \to y + z \equiv x$, and so $x \in clconv(S)$ by definition. This completes the proof. $\square$

2.21 a. The extreme points of S are defined by the intersection of the two defining constraints, which yield upon solving for x_1 and x_2 in terms of x_3 that

$$x_1 = -1 \pm \sqrt{5 - 2x_3}, \ x_2 = \frac{3 - x_3 \mp \sqrt{5 - 2x_3}}{2}, \text{ where } x_3 \le \frac{5}{2}.$$

For characterizing the extreme directions of S, first note that for any fixed x_3, we have that S is bounded. Thus, any extreme direction must have $d_3 \ne 0$. Moreover, the maximum value of x_3 over S is readily verified to be bounded. Thus, we can set $d_3 = -1$. Furthermore, if $\bar{x} \equiv (0,0,0)$ and $d = (d_1, d_2, -1)$, then $\bar{x} + \lambda d \in S$, $\forall \lambda > 0$, implies that

$$d_1 + 2d_2 \le 1 \tag{1}$$

and that $4\lambda d_2 \ge \lambda^2 d_1^2$, i.e., $4d_2 \ge \lambda^2 d_1^2$, $\forall \lambda > 0$. Hence, if $d_1 \ne 0$, then we will have $d_2 \to \infty$, and so (for bounded direction components) we must have $d_1 = 0$ and $d_2 \ge 0$. Thus together with (1), for extreme directions, we can take $d_2 = 0$ or $d_2 = 1/2$, yielding

$(0,0,-1)$ and $(0, \dfrac{1}{2}, -1)$ as the extreme directions of S.

b. Since S is a polyhedron in R^3, its extreme points are feasible solutions defined by the intersection of three linearly independent defining hyperplanes, of which one must be the equality restriction $x_1 + x_2 = 1$. Of the six possible choices of selecting two from the remaining four defining constraints, we get extreme points defined by four such choices (easily verified), which yields $(0,1,\dfrac{3}{2})$, $(1,0,\dfrac{3}{2})$, $(0,1,0)$, and $(1,0,0)$ as the four extreme points of S. The extreme directions of S are given by extreme points of $D \equiv \{(d_1, d_2, d_3) : d_1 + d_2 + 2d_3 \le 0, \ d_1 + d_2 = 0, \ d_1 + d_2 + d_3 = 1, \ d \ge 0\}$, which is empty. Thus, there are no extreme directions of S (i.e., S is bounded).

c. From a plot of S, it is readily seen that the extreme points of S are given by $(0, 0)$, plus all point on the circle boundary $x_1^2 + x_2^2 = 2$ that lie between the points $(-\sqrt{2/5}, 2\sqrt{2/5})$ and $(\sqrt{2/5}, 2\sqrt{2/5})$, including the two end-points. Furthermore, since S is bounded, it has no extreme direction.

2.24 By plotting (or examining pairs of linearly independent active constraints), we have that the extreme points of S are given by $(0, 0)$, $(3, 0)$, and $(0, 2)$. Furthermore, the extreme directions of S are given by extreme points of
$$D = \{(d_1, d_2): \quad -d_1 + 2d_2 \leq 0 \quad d_1 - 3d_2 \leq 0, \quad d_1 + d_2 = 1, \quad d \geq 0\},$$
which are readily obtained as $(\frac{2}{3}, \frac{1}{3})$ and $(\frac{3}{4}, \frac{1}{4})$. Now, let
$$\begin{bmatrix} 4 \\ 1 \end{bmatrix} = \begin{bmatrix} \bar{x}_1 \\ x_2 \end{bmatrix} + \lambda \begin{bmatrix} 3/4 \\ 1/4 \end{bmatrix}, \text{ where } \begin{bmatrix} \bar{x}_1 \\ x_2 \end{bmatrix} = \mu \begin{bmatrix} 3 \\ 0 \end{bmatrix} + (1-\mu) \begin{bmatrix} 0 \\ 2 \end{bmatrix},$$
for $(\mu, \lambda) > 0$. Solving, we get $\mu = 7/9$ and $\lambda = 20/9$, which yields
$$\begin{bmatrix} 4 \\ 1 \end{bmatrix} = \frac{7}{9} \begin{bmatrix} 3 \\ 0 \end{bmatrix} + \frac{2}{9} \begin{bmatrix} 0 \\ 2 \end{bmatrix} + \frac{20}{9} \begin{bmatrix} 3/4 \\ 1/4 \end{bmatrix}.$$

2.31 The following result from linear algebra is very useful in this proof:
(*) An $(m + 1) \times (m + 1)$ matrix G with a row of ones is invertible if and only if the remaining m rows of G are linearly independent. In other words, if $G = \begin{bmatrix} B & a \\ e^t & 1 \end{bmatrix}$, where B is an $m \times m$ matrix, a is an $m \times 1$ vector, and e is an $m \times 1$ vector of ones, then G is invertible if and only if B is invertible. Moreover, if G is invertible, then
$$G^{-1} = \begin{bmatrix} M & g \\ h^t & f \end{bmatrix}, \text{ where } M = B^{-1}(I + \frac{1}{\alpha} ae^t B^{-1}), \quad g = -\frac{1}{\alpha} B^{-1} a,$$
$h^t = -\frac{1}{\alpha} e^t B^{-1}$, and $f = \frac{1}{\alpha}$, and where $\alpha = 1 - e^t B^{-1} a$.

By Theorem 2.6.4, an n-dimensional vector d is an extreme point of D if and only if the matrix $\begin{bmatrix} A \\ e^t \end{bmatrix}$ can be decomposed into $[B_D \ N_D]$ such that $\begin{bmatrix} d_B \\ d_N \end{bmatrix}$, where $d_N = 0$ and $d_B = B_D^{-1} b_D \geq 0$, where $b_D = \begin{bmatrix} 0 \\ 1 \end{bmatrix}$. From Property (*) above, the matrix $\begin{bmatrix} A \\ e^t \end{bmatrix}$ can be decomposed into $[B_D \ N_D]$, where B_D is a nonsingular matrix, if and only if A can be decomposed into $[B \ N]$, where B is an $m \times m$ invertible matrix. Thus, the matrix B_D must

9

necessarily be of the form $\begin{bmatrix} B & a_j \\ e^t & 1 \end{bmatrix}$, where B is an $m \times m$ invertible submatrix of A. By applying the above equation for the inverse of G, we obtain

$$d_B = B_D^{-1} b_D = \begin{bmatrix} -\dfrac{1}{\alpha} B^{-1} a_j \\ \dfrac{1}{\alpha} \end{bmatrix} = \frac{1}{\alpha} \begin{bmatrix} -B^{-1} a_j \\ 1 \end{bmatrix},$$

where $\alpha = 1 - e^t B^{-1} a_j$. Notice that $d_B \geq 0$ if and only if $\alpha > 0$ and $B^{-1} a_j \leq 0$. This result, together with Theorem 2.6.6, leads to the conclusion that d is an extreme point of D if and only if d is an extreme direction of S.

Thus, for characterizing the extreme points of D, we can examine bases of $\begin{bmatrix} A \\ e^t \end{bmatrix}$, which are limited by the number of ways we can select $(m + 1)$ columns out of n, i.e.,

$$\binom{n}{m+1} = \frac{n!}{(m+1)!(n-m-1)!},$$

which is fewer by a factor of $\dfrac{1}{(m+1)}$ than that of the Corollary to Theorem 2.6.6.

2.42 Problem P: Minimize $\{c^t x : Ax = b, \ x \geq 0\}$.

(Homogeneous) Problem D: Maximize $\{b^t y : A^t y \leq 0\}$.

Problem P has no feasible solution if and only if the system $Ax = b$, $x \geq 0$, is inconsistent. That is, by Farkas' Theorem (Theorem 2.4.5), this occurs if and only if the system $A^t y \leq 0$, $b^t y > 0$ has a solution, i.e., if and only if the homogeneous version of the dual problem is unbounded. □

2.45 Consider the following pair of primal and dual LPs, where e is a vector of ones in $\mathbb{R}^m$:

P: Max $e^t p$ D: Min $0^t x$
       subject to $A^t p = 0$ $Ax \geq e$
             $p \geq 0.$ x unres.

Then, System 2 has a solution $\Leftrightarrow P$ is unbounded (take any feasible solution to System 2, multiply it by a scalar λ, and take $\lambda \to \infty$) $\Leftrightarrow D$

is infeasible (since P is homogeneous) $\Leftrightarrow$ $\nexists$ a solution to $Ax > 0$ $\Leftrightarrow$ $\nexists$ a solution to $Ax < 0$. $\square$

2.47 Consider the system $A^t y = c$, $y \geq 0$:

$$2y_1 + 2y_2 = -3$$
$$y_1 + 2y_2 = 1$$
$$-3y_1 = -2$$
$$(y_1, y_2) \geq 0.$$

The first equation is in conflict with $(y_1, y_2) \geq 0$. Therefore, this system has no solution. By Farkas' Theorem we then conclude that the system $Ax \leq 0$, $c^t x > 0$ has a solution.

2.49 ($\Rightarrow$) We show that if System 2 has a solution, then System 1 is inconsistent. Suppose that System 2 is consistent and let y_0 be its solution. If System 1 has a solution, x_0, say, then we necessarily have $x_0^t A^t y_0 = 0$. However, since $x_0^t A^t = c^t$, this result leads to $c^t y_0 = 0$, thus contradicting $c^t y_0 = 1$. Therefore, System 1 must be inconsistent.

($\Leftarrow$) In this part we show that if System 2 has no solution, then System 1 has one. Assume that System 2 has no solution, and let $S = \{(z_1, z_0):$ $z_1 = -A^t y,\ z_0 = c^t y,\ y \in \mathbb{R}^m\}$. Then S is a nonempty convex set, and $(z_1, z_0) = (0, 1) \notin S$. Therefore, there exists a nonzero vector (p_1, p_0) and a real number α such that $p_1^t z_1 + p_0 z_0 \leq \alpha < p_1^t 0 + p_0$ for any $(z_1, z_0) \in S$. By the definition of S, this implies that $-p_1^t A^t y + p_0 c^t y \leq \alpha < p_0$ for any $y \in \mathbb{R}^m$. In particular, for $y = 0$, we obtain $0 \leq \alpha < p_0$. Next, observe that since α is nonnegative and $(-p_1^t A^t + p_0 c^t) y \leq \alpha$ for any $y \in \mathbb{R}^m$, then we necessarily have $-p_1^t A^t + p_0 c^t = 0$ (or else y can be readily selected to violate this inequality). We have thus shown that there exists a vector (p_1, p_0) where $p_0 > 0$, such that $Ap_1 - p_0 c = 0$. By letting $x = \dfrac{1}{p_0} p_1$, we concluce that x solves the system $Ax - c = 0$. This shows that System 1 has a solution. $\square$

11

2.50 Consider the pair of primal and dual LPs below, where e is a vector of ones in $\mathbb{R}^p$:

P: Max $e^t u$
     subject to $A^t u + B^t v = 0$
         $u \geq 0,\ v$ unres.

D: Min $0^t x$
     subject to $Ax \geq e$
         $Bx = 0$
         x unres.

Hence, System 2 has a solution $\Leftrightarrow P$ is unbounded (take any solution to System 2 and multiply it with a scalar λ and take $\lambda \to \infty$) $\Leftrightarrow D$ is infeasible (since P is homogeneous) $\Leftrightarrow$ there does not exist a solution to $Ax > 0$, $Bx = 0$ $\Leftrightarrow$ System 1 has no solution. $\square$

2.51 Consider the following two systems for each $i \in \{1,...,m\}$:

System I: $Ax \geq 0$ with $A_i x > 0$

System II: $A^t y = 0,\ y \geq 0$, with $y_i > 0$,

where A_i is the ith row of A. Accordingly, consider the following pair of primal and dual LPs:

P: Max $e_i^t y$
     subject to $A^t y = 0$
         $y \geq 0$

D: Min $0^t x$
     subject to $Ax \geq e_i$
         x unres,

where e_i is the ith unit vector. Then, we have that System II has a solution $\Leftrightarrow P$ is unbounded $\Leftrightarrow D$ is infeasible $\Leftrightarrow$ System I has no solution. Thus, exactly one of the systems has a solution for each $i \in \{1,...,m\}$. Let $I_1 = \{i \in \{1,...,m\}:$ System I has a solution; say $x^i\}$, and let $I_2 = \{i \in \{1,...,m\}:$ System II has a solution; say, $y^i\}$. Note that $I_1 \cup I_2 = \{1,...,m\}$ with $I_1 \cap I_2 = \varnothing$. Accordingly, let $\bar{x} = \sum_{i \in I_1} x^i$ and $\bar{y} = \sum_{i \in I_2} y^i$, where $\bar{x} \equiv 0$ if $I_1 = \varnothing$ and $\bar{y} \equiv 0$ if $I_2 = \varnothing$. Then it is easily verified that $\bar{x}$ and $\bar{y}$ satisfy Systems 1 and 2, respectively, with $A\bar{x} + \bar{y} = \sum_{i \in I_1} Ax^i + \sum_{i \in I_2} y^i > 0$ since $Ax^i \geq 0$, $\forall i \in I_1$, and $y^i \geq 0$, $\forall i \in I_2$, and moreover, for each row i of this system, if $\forall i \in I_1$ then we have $A_i x^i > 0$ and if $i \in I_2$ then we have $y^i > 0$.

2.52 Let $f(x) = e^{-x_1} - x_2$. Then $S_1 = \{x : f(x) \le 0\}$. Moreover, the Hessian

of f is given by $\begin{bmatrix} e^{-x_1} & 0 \\ 0 & 0 \end{bmatrix}$, which is positive semidefinite, and so, f is a

convex function. Thus, S is a convex set since it is a lower-level set of a convex function. Similarly, it is readily verified that S_2 is a convex set.

Furthermore, if $\bar{x} \in S_1 \cap S_2$, then we have $-e^{-\bar{x}_1} \ge \bar{x}_2 \ge e^{-\bar{x}_1}$ or

$2e^{-\bar{x}_1} \le 0$, which is achieved only in the limit as $\bar{x}_1 \to \infty$. Thus,

$S_1 \cap S_2 = \varnothing$. A separating hyperplane is given by $x_2 = 0$, with

$S_1 \subseteq \{x : x_2 \ge 0\}$ and $S_2 \subseteq \{x : x_2 \le 0\}$, but there does not exist any strongly separately hyperplane (since from above, both S_1 and S_2 contain points having $x_2 \to 0$).

2.53 Let $f(x) = x_1^2 + x_2^2 - 4$. Let $X = \{\bar{x} : \bar{x}_1^2 + \bar{x}_2^2 = 4\}$. Then, for any $\bar{x} \in X$, the first-order approximation to $f(x)$ is given by

$$f_{FO}(x) = f(\bar{x}) + (x - \bar{x})^t \nabla f(\bar{x}) = (x - \bar{x})^t \begin{bmatrix} 2\bar{x}_1 \\ 2\bar{x}_2 \end{bmatrix} = (2\bar{x}_1)x_1 + (2\bar{x}_2)x_2 - 8.$$

Thus S is described by the intersection of infinite halfspaces as follows:

$$(2\bar{x}_1)x_1 + (2\bar{x}_2)x_2 \le 8, \ \forall \bar{x} \in X,$$

which represents replacing the constraint defining S by its first-order approximation at all boundary points.

2.57 For the existence and uniqueness proof see, for example, *Linear Algebra and Its Applications* by Gilbert Strang (Harcourt Brace Jovanovich, Inc., 1988).

If $L = \{(x_1, x_2, x_3) : 2x_1 + x_2 - x_3 = 0\}$, then L is the nullspace of

$A = [2 \ 1 \ -1]$, and its orthogonal complement is given by $\lambda \begin{bmatrix} 2 \\ 1 \\ -1 \end{bmatrix}$ for any

$\lambda \in \mathbb{R}$. Therefore, $\mathbf{x}_1$ and $\mathbf{x}_2$ are orthogonal projections of $\mathbf{x}$ onto L, and

$L^\perp$, respectively. If $\mathbf{x} = (1 \ \ 2 \ \ 3)$, then $\begin{bmatrix} 1 \\ 2 \\ 3 \end{bmatrix} = \mathbf{x}_1 + \mathbf{x}_2$ where $\mathbf{x}_2 = \lambda \begin{bmatrix} 2 \\ 1 \\ -1 \end{bmatrix}$.

Thus, $\begin{bmatrix} 1 \\ 2 \\ 3 \end{bmatrix}^t \begin{bmatrix} 2 \\ 1 \\ -1 \end{bmatrix} = \lambda \left\| \begin{matrix} 2 \\ 1 \\ -1 \end{matrix} \right\|^2 \Rightarrow \lambda = \dfrac{1}{6}.$ Hence, $\mathbf{x}_1 = \dfrac{1}{6}(4 \ 11 \ 19)$ and

$\mathbf{x}_2 = \dfrac{1}{6}(2 \ 1 \ -1).$

CHAPTER 3

CONVEX FUNCTIONS AND GENERALIZATIONS

3.1 a. $\begin{bmatrix} 4 & -4 \\ -4 & 0 \end{bmatrix}$ is indefinite. Therefore, $f(x)$ is neither convex nor concave.

b. $H(x) = e^{-(x_1 + 3x_2)} \begin{bmatrix} x_1 - 2 & 3(x_1 - 1) \\ 3(x_1 - 1) & 9x_1 \end{bmatrix}$. Definiteness of the matrix $H(x)$ depends on x_1. Therefore, $f(x)$ is neither convex nor concave (over R^2).

c. $H = \begin{bmatrix} -2 & 4 \\ 4 & -6 \end{bmatrix}$ is indefinite since the determinant is negative. Therefore, $f(x)$ is neither convex nor concave.

d. $H = \begin{bmatrix} 4 & 2 & -5 \\ 2 & 2 & 0 \\ -5 & 0 & 4 \end{bmatrix}$ is indefinite. Therefore, $f(x)$ is neither convex nor concave.

e. $H = \begin{bmatrix} -4 & 8 & 3 \\ 8 & -6 & 4 \\ 3 & 4 & -4 \end{bmatrix}$ is indefinite. Therefore, $f(x)$ is neither convex nor concave.

3.2 $f''(x) = abx^{b-2} e^{-ax^b} [abx^b - (b-1)]$. Hence, if $b = 1$, then f is convex over $\{x : x > 0\}$. If $b > 1$, then f is convex whenever $abx^b \geq (b-1)$, i.e., $x \geq \left[\dfrac{(b-1)}{ab} \right]^{1/b}$.

3.3 $f(x) = 10 - 3(x_2 - x_1^2)^2$, and its Hessian matrix is $H(x) = 6 \begin{bmatrix} -6x_1^2 + 2x_2 & 2x_1 \\ 2x_1 & -1 \end{bmatrix}$. Thus, f is not convex anywhere and for f to be concave, we need $-6x_1^2 + 2x_2 \leq 0$ and $6x_1^2 - 2x_2 - 4x_1^2 \geq 0$, i.e., $3x_1^2 \geq x_2$ and $x_1^2 \geq x_2$, i.e., $x_1^2 \geq x_2$. Hence, if $S = \{(x_1, x_2) : -1 \leq x_1 \leq 1, -1 \leq x_2 \leq 1\}$, then $f(x)$ is neither convex nor concave on S.

If S is a convex set such that $S \subseteq \{(x_1, x_2) : x_1^2 \geq x_2\}$, then $H(x)$ is negative semidefinite for all $x \in S$. Therefore, $f(x)$ is concave on S.

3.4 $f(x) = x^2(x^2 - 1)$, $f'(x) = 4x^3 - 2x$, and $f''(x) = 12x^2 - 2 \geq 0$ if $x^2 \geq 1/6$. Thus f is convex over $S_1 = \{x : x \geq 1/\sqrt{6}\}$ and over $S_2 = \{x : x \leq -1/\sqrt{6}\}$. Moreover, since $f''(x) > 0$ whenever $x > 1/\sqrt{6}$ or $x < -1/\sqrt{6}$, and thus f lies strictly above the tangent plane for all $x \in S_1$ as well as for all $x \in S_2$, f is strictly convex over S_1 and over S_2. For all the remaining values for x, $f(x)$ is strictly concave.

3.9 Consider any x_1, $x_2 \in R^n$, and let $x_\lambda = \lambda x_1 + (1 - \lambda)x_2$ for any $0 \leq \lambda \leq 1$. Then

$f(x_\lambda) = \max\{f_1(x_\lambda), ..., f_k(x_\lambda)\} = f_r(x_\lambda)$ for some $r \in \{1, ..., k\}$, whence $f_r(x_\lambda) \leq \lambda f_r(x_1) + (1 - \lambda)f_r(x_2)$ by the convexity of f_r, i.e., $f(x_\lambda) \leq \lambda f(x_1) + (1 - \lambda)f(x_2)$ since $f(x_1) \geq f_r(x_1)$ and $f(x_2) \geq f_r(x_2)$. Thus f is convex.

If $f_1, ..., f_k$ are concave functions, then $-f_1, ..., -f_k$ are convex functions $\Rightarrow \max\{-f_1(x), ..., -f_k(x)\}$ is convex i.e., $-\min\{f_1(x), ..., f_k(x)\}$ is convex, i.e., $f(x) \equiv \min\{f_1(x), ..., f_k(x)\}$ is concave.

3.10 Let x_1, $x_2 \in \mathbb{R}^n$, $\lambda \in [0,1]$, and let $x_\lambda = \lambda x_1 + (1 - \lambda)x_2$. To establish the convexity of $f(\cdot)$ we need to show that $f(x_\lambda) \leq \lambda f(x_1) + (1 - \lambda)f(x_2)$. Notice that
$$f(x_\lambda) = g[h(x_\lambda)] \leq g[\lambda h(x_1) + (1 - \lambda)h(x_2)]$$
$$\leq \lambda g[h(x_1)] + (1 - \lambda)g[h(x_2)]$$
$$= \lambda f(x_1) + (1 - \lambda)f(x_2).$$
In this derivation, the first inequality follows since h is convex and g is nondecreasing, and the second inequality follows from the convexity of g. This completes the proof.

3.11 Let x_1, $x_2 \in S$, $\lambda \in [0,1]$, and let $x_\lambda = \lambda x_1 + (1 - \lambda)x_2$. To establish the convexity of f over S we need to show that $f(x_\lambda) - \lambda f(x_1) - (1 - \lambda)f(x_2) \leq 0$. For notational convenience, let

16

$D(x) = g(x_1)g(x_2) - \lambda g(x_\lambda)g(x_2) - (1 - \lambda)g(x_\lambda)g(x_2)$.   Under   the
assumption that $g(x) > 0$ for all $x \in S$, our task reduces to demonstrating
that $D(x) \leq 0$ for any x_1, $x_2 \in S$, and any $\lambda \in [0,1]$. By the concavity of
$g(x)$ we have

$D(x) \leq g(x_1)g(x_2) - \lambda[\lambda g(x_1) + (1 - \lambda)g(x_2)]g(x_2) -$
$$(1 - \lambda)[\lambda g(x_1) + (1 - \lambda)g(x_2)]g(x_1).$$

After a rearrangement of terms on the right-hand side of this inequality we
obtain

$$D(x) \leq -\lambda(1 - \lambda)[g(x_1)^2 + g(x_2)^2] + 2\lambda(1 - \lambda)g(x_1)g(x_2)$$
$$= -\lambda(1 - \lambda)[g(x_1)^2 + g(x_2)^2] + 2\lambda(1 - \lambda)g(x_1)g(x_2)$$
$$= -\lambda(1 - \lambda)[g(x_1)^2 + g(x_2)^2 - 2g(x_1)g(x_2)]$$
$$= -\lambda(1 - \lambda)[g(x_1) - g(x_2)]^2.$$

Therefore, $D(x) \leq 0$ for any x_1, $x_2 \in S$, and any $\lambda \in [0,1]$, and thus
$f(x)$ is a convex function.

Symmetrically, if g is convex, $S = \{x : g(x) < 0\}$, then from above, $\dfrac{1}{-g}$

is convex over S, and so $f(x) = 1/g(x)$ is concave over S.   □

3.16 Let x_1, x_2 be any two vectors in R^n, and let $\lambda \in [0,1]$. Then, by the
definition   of $h(\cdot)$,   we   obtain $h(\lambda x_1 + (1 - \lambda)x_2) = \lambda(Ax_1 + b) +$
$(1 - \lambda)(Ax_2 + b) = \lambda h(x_1) + (1 - \lambda)h(x_2)$. Therefore,

$f(\lambda x_1 + (1 - \lambda)x_2) = g[h(\lambda x_1 + (1 - \lambda)x_2)] = g[\lambda h(x_1) + (1 - \lambda)h(x_2)]$
$\leq \lambda g[h(x_1)] + (1 - \lambda)g[h(x_2)] = \lambda f(x_1) + (1 - \lambda)f(x_2)$,

where the above inequality follows from the convexity of g. Hence, $f(x)$
is convex.   □

By multivariate calculus, we obtain $\nabla f(x) = A^t \nabla g[h(x)]$, and $H_f(x) = $
$A^t H_g[h(x)]A$.

3.18 Assume that $f(x)$ is convex. Consider any x, $y \in R^n$, and let $\lambda \in (0,1)$.
Then

$$f(x + y) = f\left[\lambda\left(\frac{x}{\lambda}\right) + (1 - \lambda)\left(\frac{y}{1 - \lambda}\right)\right] \leq \lambda f\left(\frac{x}{\lambda}\right) + (1 - \lambda)f\left(\frac{y}{1 - \lambda}\right)$$

$$= f(x) + f(y),$$
and so f is subadditive.

Conversely, let f be a subadditive gauge function. Let x, $y \in R^n$ and $\lambda \in [0,1]$. Then
$$f(\lambda x + (1-\lambda)y) \le f(\lambda x) + f[(1-\lambda)y] = \lambda f(x) + (1-\lambda)f(y),$$
and so f is convex.

3.21 See the answer to Exercise 6.4.

3.22 a. See the answer to Exercise 6.4.

b. If $y_1 \le y_2$, then $\{x : g(x) \le y_1, x \in S\} \subseteq \{x : g(x) \le y_2, x \in S\}$, and so $\phi(y_1) \ge \phi(y_2)$.

3.26 First assume that $\bar{x} = 0$. Note that then $f(\bar{x}) = 0$ and $\xi^t \bar{x} = 0$ for any vector ξ in R^n.

$(\Rightarrow)$ If ξ is a subgradient of $f(x) = \|x\|$ at $x = 0$, then by definition we have $\|x\| \ge \xi^t x$ for all $x \in R^n$. Thus in particular for $x = \xi$, we obtain $\|\xi\| \ge \|\xi\|^2$, which yields $\|\xi\| \le 1$.

$(\Leftarrow)$ Suppose that $\|\xi\| \le 1$. By the Schwarz inequality, we then obtain $\xi^t x \le \|\xi\| \, \|x\| \le \|x\|$, and so ξ is a subgradient of $f(x) = \|x\|$ at $x = 0$.
This completes the proof for the case when $\bar{x} = 0$. Now, consider $\bar{x} \ne 0$.

$(\Rightarrow)$ Suppose that ξ is a subgradient of $f(x) = \|x\|$ at $\bar{x}$. Then by definition, we have

$$\|x\| - \|\bar{x}\| \ge \xi^t (x - \bar{x}) \text{ for all } x \in R^n. \tag{1}$$

In particular, the above inequality holds for $x = 0$, for $x = \lambda \bar{x}$, where $\lambda > 0$, and for $x = \xi$. If $x = 0$, then $\xi^t \bar{x} \ge \|\bar{x}\|$. Furthermore, by employing the Schwarz inequality we obtain

$$\|\bar{x}\| \le \xi^t \bar{x} \le \|\xi\| \, \|\bar{x}\|. \tag{2}$$

If $x = \lambda \bar{x}$, $\lambda > 0$, then $\|x\| = \lambda \|\bar{x}\|$, and Equation (1) yields $(\lambda - 1)\|\bar{x}\| \ge (\lambda - 1)\xi^t \bar{x}$. If $\lambda > 1$, then $\|\bar{x}\| \ge \xi^t \bar{x}$, and if $\lambda < 1$, then

$\|\bar{x}\| \leq \xi^t \bar{x}$. Therefore, in either case, if ξ is a subgradient at $\bar{x}$, then it must satisfy the equation.

$$\xi^t \bar{x} = \|\bar{x}\|.$$ (3)

Finally, if $x = \xi$, then Equation (1) results in $\|\xi\| - \|\bar{x}\| \geq \xi^t \xi - \xi^t \bar{x}$. However, by (2), we have $\xi^t \bar{x} = \|\bar{x}\|$. Therefore, $\|\xi\|(1 - \|\xi\|) \geq 0$. This yields

$$1 - \|\xi\| \geq 0$$ (4)

Combining (2) – (4), we conclude that if ξ is a subgradient of $f(x) = \|x\|$ at $\bar{x} \neq 0$, then $\xi^t \bar{x} = \|\bar{x}\|$ and $\|\xi\| = 1$.

($\Leftarrow$) Consider a vector $\xi \in R^n$ such that $\|\xi\| = 1$ and $\xi^t \bar{x} = \|\bar{x}\|$, where $\bar{x} \neq 0$. Then for any x, we have $f(x) - f(\bar{x}) - \xi^t(x - \bar{x}) = \|x\| - \|\bar{x}\| - \xi^t(x - \bar{x}) = \|x\| - \xi^t x \geq \|x\|(1 - \|\xi\|) = 0$, where we have used the Schwarz inequality ($\xi^t x \leq \|\xi\| \|x\|$) to derive the last inequality. Thus ξ is a subgradient of $f(x) = \|x\|$ at $\bar{x} \neq 0$. This completes the proof. $\square$

In order to derive the gradient of $f(x)$ at $\bar{x} \neq 0$, notice that $\|\xi\| = 1$ and $\xi^t \bar{x} = \|\bar{x}\|$ if and only if $\xi = \frac{1}{\|\bar{x}\|} \bar{x}$. Thus $\nabla f(\bar{x}) = \frac{1}{\|\bar{x}\|} \bar{x}$.

3.27 Since f_1 and f_2 are convex and differentiable, we have

$$f_1(x) \geq f_1(\bar{x}) + (x - \bar{x})^t \nabla f_1(\bar{x}), \quad \forall x.$$
$$f_2(x) \geq f_2(\bar{x}) + (x - \bar{x})^t \nabla f_2(\bar{x}), \quad \forall x.$$

Hence, $f(x) = \max\{f_1(x), f_2(x)\}$ and $f(\bar{x}) = f_1(\bar{x}) = f_2(\bar{x})$ give

$$f(x) \geq f(\bar{x}) + (x - \bar{x})^t \nabla f_1(\bar{x}), \quad \forall x$$ (1)
$$f(x) \geq f(\bar{x}) + (x - \bar{x})^t \nabla f_2(\bar{x}), \quad \forall x.$$ (2)

Multiplying (1) and (2) by λ and $(1 - \lambda)$, respectively, where $0 \leq \lambda \leq 1$, yields upon summing:

$$f(x) \geq f(\bar{x}) + (x - \bar{x})'[\lambda \nabla f_1(\bar{x}) + (1 - \lambda)\nabla f_2(\bar{x})], \quad \forall x,$$

$\Rightarrow \quad \xi = \lambda \nabla f_1(\bar{x}) + (1 - \lambda)\nabla f_2(\bar{x}), \; 0 \leq \lambda \leq 1$, is a subgradient of f at $\bar{x}$.

$(\Rightarrow)$ Let ξ be a subgradient of f at $\bar{x}$. Then, we have,

$$f(x) \geq f(\bar{x}) + (x - \bar{x})' \xi, \quad \forall x. \tag{3}$$

But $f(x) = \max\{f_1(x), f_2(x)\} =$

$$\max\{f_1(\bar{x}) + (x - \bar{x})' \nabla f_1(\bar{x}) + \|x - \bar{x}\| 0_1(x \to \bar{x}),$$
$$f_2(\bar{x}) + (x - \bar{x})' \nabla f_2(\bar{x}) + \|x - \bar{x}\| 0_2(x \to \bar{x})\}, \tag{4}$$

where $0_1(x \to \bar{x})$ and $0_2(x \to \bar{x})$ are functions that approach zero as $x \to \bar{x}$. Since $f_1(\bar{x}) = f_2(\bar{x}) = f(\bar{x})$, putting (3) and (4) together yields

$$\max\{(x - \bar{x})'[\nabla f_1(\bar{x}) - \xi] + \|x - \bar{x}\| 0_1(x \to \bar{x}),$$
$$(x - \bar{x})'[\nabla f_2(\bar{x}) - \xi] + \|x - \bar{x}\| 0_2(x \to \bar{x})\} \geq 0, \quad \forall x. \tag{5}$$

Now, on the contrary, suppose that $\xi \notin conv\{\nabla f_1(\bar{x}), \nabla f_2(\bar{x})\}$. Then, there exists a strictly separating hyperplane $\alpha x = \beta$ such that $\|\alpha\| = 1$ and $\alpha' \xi > \beta$ and $\{\alpha' \nabla f_1(\bar{x}) < \beta, \; \alpha' \nabla f_2(\bar{x}) < \beta\}$, i.e.,

$$\alpha'[\xi - \nabla f_1(\bar{x})] > 0 \text{ and } \alpha'[\xi - \nabla f_2(\bar{x})] > 0. \tag{6}$$

Letting $(x - \bar{x}) = \varepsilon \alpha$ in (5), with $\varepsilon \to 0^+$, we get upon dividing with $\varepsilon > 0$:

$$\max\{\alpha'[\nabla f_1(\bar{x}) - \xi] + 0_1(\varepsilon \to 0),$$
$$\alpha'[\nabla f_2(\bar{x}) - \xi] + 0_2(\varepsilon \to 0)\} \geq 0, \quad \forall \varepsilon > 0. \tag{7}$$

But the first terms in both maxands in (7) are negative by (6), while the second terms $\to 0$. Hence we get a contradiction. Thus $\xi \in conv\{\nabla f_1(\bar{x}), \nabla f_2(\bar{x})\}$, i.e., it is of the given form.

Similarly, if $f(x) = \max\{f_1(x), ..., f_m(x)\}$, where $f_1, ..., f_m$ are differentiable convex functions and $\bar{x}$ is such that $f(\bar{x}) = f_i(\bar{x})$, $\forall i \in I \subseteq \{1, ..., m\}$, then ξ is a subgradient of f at $\bar{x} \Leftrightarrow \xi \in conv\{\nabla f_i(\bar{x}), i \in I\}$. A likewise result holds for the minimum of differentiable concave functions.

3.28 a. See Theorem 6.3.1 and its proof. (Alternatively, since θ is the minimum of several affine functions, one for each extreme point of X, we have that θ is a piecewise linear and concave.)

b. See Theorem 6.3.7. In particular, for a given vector $\bar{u}$, let $X(\bar{u}) = \{x_1, ..., x_k\}$ denote the set of all extreme points of the set X that are optimal solutions for the problem to minimize $\{c^t x + \bar{u}^t(Ax - b) : x \in X\}$. Then $\xi(\bar{u})$ is a subgradient of $\theta(u)$ at $\bar{u}$ if and only if $\xi(\bar{u})$ is in the convex hull of $Ax_1 - b, ..., Ax_k - b$, where $x_i \in X(\bar{u})$ for $i = 1, ..., k$. That is, $\xi(\bar{u})$ is a subgradient of

$$\theta(u) \text{ at } \bar{u} \text{ if and only if } \xi(\bar{u}) = A\sum_{i=1}^{k} \lambda_i x_i - b \text{ for some nonnegative}$$

$\lambda_1, ..., \lambda_k$, such that $\sum_{i=1}^{k} \lambda_i = 1$.

3.31 Let $P_1 : \min\{f(x) : x \in S\}$ and $P_2 : \min\{f_s(x) : x \in S\}$, and let $S_1 = \{x^* \in S : f(x^*) \le f(x), \forall x \in S\}$ and $S_2 = \{x^* \in S : f_s(x^*) \le f_s(x), \forall x \in S\}$. Consider any $x^* \in S_1$. Hence, x^* solves Problem P_1. Define $h(x) = f(x^*), \forall x \in S$. Thus, the constant function h is a convex underestimating function for f over S, and so by the definition of f_s, we have that

$$f_s(x) \ge h(x) = f(x^*), \forall x \in S. \tag{1}$$

But $f_s(x^*) \le f(x^*)$ since $f_s(x) \le f(x), \forall x \in S$. This, together with (1), thus yields $f_s(x^*) = f(x^*)$ and that x^* solves Problem P_2 (since (1) asserts that $f(x^*)$ is a lower bound on Problem P_2). Therefore, $x^* \in S_2$. Thus, we have shown that the optimal values of Problems P_1 and P_2 match, and that $S_1 \subseteq S_2$. $\square$

21

3.37 $\nabla f(x) = \begin{bmatrix} 4x_1 e^{2x_1^2 - x_2^2} & -3 \\ -2x_2 e^{2x_1^2 - x_2^2} & +5 \end{bmatrix}$, $\nabla f\begin{bmatrix} 1 \\ 1 \end{bmatrix} = \begin{bmatrix} 4e - 3 \\ -2e + 5 \end{bmatrix}$

$H(x) = 2e^{2x_1^2 - x_2^2} \begin{bmatrix} 8x_1^2 + 2 & -4x_1 x_2 \\ -4x_1 x_2 & 2x_2^2 - 1 \end{bmatrix}$, $H\begin{bmatrix} 1 \\ 1 \end{bmatrix} = 2e\begin{bmatrix} 10 & -4 \\ -4 & 1 \end{bmatrix}$,

with $f\begin{bmatrix} 1 \\ 1 \end{bmatrix} = e + 2$.

Thus, the linear (first-order) approximation of f at $\begin{bmatrix} 1 \\ 1 \end{bmatrix}$ is given by

$f_1(x) \equiv (e + 2) + (x_1 - 1)(4e - 3) + (x_2 - 1)(-2e + 5)$,

and the second-order approximation of f at $\begin{bmatrix} 1 \\ 1 \end{bmatrix}$ is given by

$f_2(x) \equiv (e + 2) + (x_1 - 1)(4e - 3) + (x_2 - 1)(-2e + 5) +$
$\qquad e\left[10(x_1 - 1)^2 - 8(x_1 - 1)(x_2 - 1) + (x_2 - 1)^2 \right]$.

f_1 is both convex and concave (since it is affine). The Hessian of f_2 is given by $H\begin{bmatrix} 1 \\ 1 \end{bmatrix}$, which is indefinite, and so f_2 is neither convex nor concave.

3.39 The function $f(x) = x^t Ax$ can be represented in a more convenient form as $f(x) = \frac{1}{2} x^t (A + A^t)x$, where $(A + A^t)$ is symmetric. Hence, the Hessian matrix of $f(x)$ is $H = A + A^t$. By the superdiagonalization procedure, we can readily verify that $H = \begin{bmatrix} 4 & 3 & 4 \\ 3 & 6 & 3 \\ 4 & 3 & 2\theta \end{bmatrix}$. H is positive semidefinite if and only if $\theta \geq 2$, and is positive definite for $\theta > 2$. Therefore, if $\theta > 2$, then $f(x)$ is strictly convex. To examine the case when $\theta = 2$, consider the following three points: $x_1 = (1, 0, 0)$, $x_2 = (0, 0, 1)$, and $\bar{x} = \frac{1}{2}x_1 + \frac{1}{2}x_2$. As a result of direct substitution, we obtain $f(x_1) = f(x_2) = 2$, and $f(\bar{x}) = 2$. This shows that $f(x)$ is not strictly convex (although it is still convex) when $\theta = 2$.

22

3.40 $f(x) = x^3 \Rightarrow f'(x) = 3x^2$ and $f''(x) = 6x \geq 0$, $\forall x \in S$. Hence f is convex on S. Moreover, $f''(x) > 0$, $\forall x \in int(S)$, and so f is strictly convex on $int(S)$. To show that f is strictly convex on S, note that $f''(x) = 0$ only for $x = 0 \in S$, and so following the argument given after Theorem 3.3.8, any supporting hyperplane to the epigraph of f over S at any point $\bar{x}$ must touch it only at $[\bar{x}, f(\bar{x})]$, or else this would contradict the strict convexity of f over $int(S)$. Note that the first nonzero derivative of order greater than or equal to 2 at $\bar{x} = 0$ is $f'''(\bar{x}) = 6$, but Theorem 3.3.9 does not apply here since $\bar{x} = 0 \in \partial(S)$. Indeed, this shows that $f(x) = x^3$ is neither convex nor concave over R. But Theorem 3.3.9 applies (and holds) over $int(S)$ in this case.

3.41 The matrix H is symmetric, and therefore, it is diagonalizable. That is, there exists an orthogonal $n \times n$ matrix Q, and a diagonal $n \times n$ matrix D such that $H = QDQ^t$. The columns of the matrix Q are simply normalized eigenvectors of the matrix H, and the diagonal elements of the matrix D are the eigenvalues of H. By the positive semidefiniteness of H, we have $diag\{D\} \geq 0$, and hence there exists a square root matrix $D^{1/2}$ of D (that is $D = D^{1/2}D^{1/2}$).

If $x = 0$, then readily $Hx = 0$. Suppose that $x^t Hx = 0$ for some $x \neq 0$. Below we show that then Hx is necessarily 0. For notational convenience let $z = D^{1/2}Q^t x$. Then the following equations are equivalent to $x^t Hx = 0$:

$$x^t Q D^{1/2} D^{1/2} Q^t x = 0$$
$$\Leftrightarrow \quad z^t z = 0, \text{ i.e., } \|z\|^2 = 0$$
$$\Leftrightarrow \quad z = 0.$$

By premultiplying the last equation by $QD^{1/2}$, we obtain $QD^{1/2}z = 0$, which by the definition of z gives $QDQ^t x = 0$. Thus $Hx = 0$, which completes the proof. $\square$

3.45 Consider the problem

> **P:** Minimize $(x_1 - 4)^2 + (x_2 - 6)^2$
> subject to $x_2 \geq x_1^2$
> $x_2 \leq 4$.

Note that the feasible region (denote this by X) of Problem P is convex. Hence, a necessary condition for $\bar{x} \in X$ to be an optimal solution for Problem P is that

$$\nabla f(\bar{x})^t(x - \bar{x}) \geq 0, \quad \forall x \in X, \tag{1}$$

because if there exists an $\hat{x} \in X$ such that $\nabla f(\bar{x})^t(\hat{x} - \bar{x}) < 0$, then $d \equiv (\hat{x} - \bar{x})$ would be an improving (since f is differentiable) and feasible (since X is convex) direction.

For $\bar{x} = (2,4)^t$, we have $\nabla f(\bar{x}) = \begin{bmatrix} 2(2-4) \\ 2(4-6) \end{bmatrix} = \begin{bmatrix} -4 \\ -4 \end{bmatrix}$.

Hence,

$$\nabla f(\bar{x})^t(x - \bar{x}) = [-4, -4] = \begin{bmatrix} x_1 - 2 \\ x_2 - 4 \end{bmatrix} = -4x_1 - 4x_2 + 24. \tag{2}$$

But $x_1^2 \leq x_2 \leq 4$, $\forall x \in X \Rightarrow x_2 \leq 4$ and $-2 \leq x_1 \leq 2$, and so $-4x_1 \geq -8$ and $-4x_2 \geq -16$. Hence, $\nabla f(\bar{x})^t(x - \bar{x}) \geq 0$ from (2).

Furthermore, observe that the objective function of Problem P (denoted by $f(x)$) is (strictly) convex since its Hessian is given by $\begin{bmatrix} 2 & 0 \\ 0 & 2 \end{bmatrix}$, which is positive definite. Hence, by Corollary 2 to Theorem 3.4.3, we have that (1) is also sufficient for optimality to P, and so $\bar{x} = (2,4)^t$ (uniquely) solves Problem P.

3.48 Suppose that λ_1 and λ_2 are in the interval $(0, \delta)$, and such that $\lambda_2 > \lambda_1$. We need to show that $f(x + \lambda_2 d) \geq f(x + \lambda_1 d)$.

Let $\alpha = \lambda_1/\lambda_2$. Note that $\alpha \in (0,1)$, and $x + \lambda_1 d = \alpha(x + \lambda_2 d) + (1 - \alpha)x$. Therefore, by the convexity of f, we obtain $f(x + \lambda_1 d) \leq \alpha f(x + \lambda_2 d) + (1 - \alpha)f(x)$, which leads to $f(x + \lambda_1 d) \leq f(x + \lambda_2 d)$ since, by assumption, $f(x) \leq f(x + \lambda d)$ for any $\lambda \in (0, \delta)$.

24

When f is strictly convex, we can simply replace the weak inequalities above with strict inequalities to conclude that $f(x + \lambda d)$ is strictly increasing over the interval $(0, \delta)$.

3.51 ($\Leftrightarrow$) If the vector d is a descent direction of f at $\bar{x}$, then $f(\bar{x} + \lambda d) - f(\bar{x}) < 0$ for all $\lambda \in (0, \delta)$. Moreover, since f is a convex and differentiable function, we have that $f(\bar{x} + \lambda d) - f(\bar{x}) \geq \lambda \nabla f(\bar{x})^t d$. Therefore, $\nabla f(\bar{x})^t d < 0$.

($\Leftrightarrow$) See the proof of Theorem 4.1.2. $\square$

Note: If the function $f(x)$ is not convex, then it is not true that $\nabla f(\bar{x})^t d < 0$ whenever d is a descent direction of $f(x)$ at $\bar{x}$. For example, if $f(x) = x^3$, then $d = -1$ is a descent direction of f at $\bar{x} = 0$, but $f'(\bar{x})d = 0$.

3.54 ($\Rightarrow$) If $\bar{x}$ is an optimal solution, then we must have $f'(\bar{x}; d) \geq 0$, $\forall d \in D$, since $f'(\bar{x}; d) < 0$ for any $d \in D$ implies the existence of improving feasible solutions by Exercise 3.5.1.

($\Leftarrow$) Suppose $f'(\bar{x}; d) \geq 0$, $\forall d \in D$, but on the contrary, $\bar{x}$ is not an optimal solution, i.e., there exists $\hat{x} \in S$ with $f(\hat{x}) < f(\bar{x})$. Consider $d = (\hat{x} - \bar{x})$. Then $d \in D$ since S is convex. Moreover, $f(\bar{x} + \lambda d) = f(\lambda \hat{x} + (1 - \lambda)\bar{x}) \leq \lambda f(\hat{x}) + (1 - \lambda)f(\bar{x}) < f(\bar{x})$, $\forall 0 < \lambda \leq 1$. Thus d is a feasible, descent direction, and so $f'(\bar{x}; d) < 0$ by Exercise 3.51, a contradiction.

Theorem 3.4.3 similarly deals with nondifferentiable convex functions.

If $S = R^n$, then $\bar{x}$ is optimal $\Leftrightarrow$ $\nabla f(\bar{x})^t d \geq 0$, $\forall d \in R^n$ $\Leftrightarrow$ $\nabla f(\bar{x}) = 0$ (else, pick $d = -\nabla f(\bar{x})$ to get a contradiction).

3.56 Let x_1, $x_2 \in R^n$. Without loss of generality assume that $h(x_1) \geq h(x_2)$. Since the function g is nondecreasing, the foregoing assumption implies that $g[h(x_1)] \geq g[h(x_2)]$, or equivalently, that $f(x_1) \geq f(x_2)$. By the quasiconvexity of h, we have $h(\alpha x_1 + (1 - \alpha)x_2) \leq h(x_1)$ for any $\alpha \in [0,1]$. Since the function g is nondecreasing, we therefore have, $f(\alpha x_1 + (1 - \alpha)x_2) = g[h(\alpha x_1 + (1 - \alpha)x_2)] \leq g[h(x_1)] = f(x_1)$. This shows that $f(x)$ is quasiconvex. $\square$

3.61 Let α be an arbitrary real number, and let $S = \{x : f(x) \le \alpha\}$. Furthermore, let x_1 and x_2 be any two elements of S. By Theorem 3.5.2, we need to show that S is a convex set, that is, $f(\lambda x_1 + (1 - \lambda)x_2) \le \alpha$ for any $\lambda \in [0,1]$. By the definition of $f(x)$, we have

$$f(\lambda x_1 + (1 - \lambda)x_2) = \frac{g(\lambda x_1 + (1 - \lambda)x_2)}{h(\lambda x_1 + (1 - \lambda)x_2)} \le \frac{\lambda g(x_1) + (1 - \lambda)g(x_2)}{\lambda h(x_1) + (1 - \lambda)h(x_2)}, \tag{1}$$

where the inequality follows from the assumed properties of the functions g and h. Furthermore, since $f(x_1) \le \alpha$ and $f(x_2) \le \alpha$, we obtain

$$\lambda g(x_1) \le \lambda \alpha h(x_1) \text{ and } (1 - \lambda)g(x_2) \le (1 - \lambda)\alpha h(x_2).$$

By adding these two inequalities, we obtain $\lambda g(x_1) + (1 - \lambda)g(x_2) \le \alpha[\lambda h(x_1) + (1 - \lambda)h(x_2)]$. Since h is assumed to be a positive-valued function, the last inequality yields

$$\frac{\lambda g(x_1) + (1 - \lambda)g(x_2)}{\lambda h(x_1) + (1 - \lambda)h(x_2)} \le \alpha,$$

or by (1), $f(\lambda x_1 + (1 - \lambda)x_2) \le \alpha$. Thus, S is a convex set, and therefore, $f(x)$ is a quasiconvex function. □

Alternative proof: For any $\alpha \in R$, let $S_\alpha = \{x \in S : g(x)/h(x) \le \alpha\}$. We need to show that S_α is a convex set. If $\alpha < 0$, then $S_\alpha = \varnothing$ since $g(x) \ge 0$ and $h(x) \ge 0$, $\forall x \in S$, and so S_α is convex. If $\alpha \ge 0$, then $S_\alpha = \{x \in S : g(x) - \alpha h(x) \le 0\}$ is convex since $g(x) - \alpha h(x)$ is a convex function, and S_α is a lower level set of this function. □

3.62 We need to prove that if $g(x)$ is a convex nonpositive-valued function on S and $h(x)$ is a convex and positive-valued function on S, then $f(x) = g(x)/h(x)$ is a quasiconvex function on S. For this purpose we show that for any x_1, $x_2 \in S$, if $f(x_1) \ge f(x_2)$, then $f(x_\lambda) \le f(x_1)$, where $x_\lambda = \lambda x_1 + (1 - \lambda)x_2$, and $\lambda \in [0,1]$. Note that by the definition of f and the assumption that $h(x) > 0$ for all $x \in S$, it suffices to show that $g(x_\lambda)h(x_1) - g(x_1)h(x_\lambda) \le 0$. Towards this end, observe that

$g(x_\lambda)h(x_1) \le [\lambda g(x_1) + (1 - \lambda)g(x_2)]h(x_1)$ since $g(x)$ is convex and $h(x) > 0$ on S;

$g(x_1)h(x_\lambda) \ge g(x_1)[\lambda h(x_1) + (1 - \lambda)h(x_2)]$ since $h(x)$ is convex and $g(x) \le 0$ on S;

$g(x_2)h(x_1) - g(x_1)h(x_2) \le 0$, since $f(x_1) \ge f(x_2)$ and $h(x) > 0$ on S.

From the foregoing inequalities we obtain

$g(x_\lambda)h(x_1) - g(x_1)h(x_\lambda)$
$\le [\lambda g(x_1) + (1 - \lambda)g(x_2)]h(x_1) - g(x_1)[\lambda h(x_1) + (1 - \lambda)h(x_2)]$
$= (1 - \lambda)[g(x_2)h(x_1) - g(x_1)h(x_2)] \le 0,$

which implies that $f(x_\lambda) \le \max\{f(x_1), f(x_2)\} = f(x_1)$. $\quad\square$

Note: See also the alternative proof technique for Exercise 3.61 for a similar simpler proof of this result.

3.63 By assumption, $h(x) \ne 0$, and so the function $f(x)$ can be rewritten as $f(x) = g(x)/p(x)$, where $p(x) \equiv 1/h(x)$. Furthermore, since $h(x)$ is a concave and positive-valued function, we conclude that $p(x)$ is convex and positive-valued on S (see Exercise 3.11). Therefore, the result given in Exercise 3.62 applies. This completes the proof. $\quad\square$

3.64 Let us show that if $g(x)$ and $h(x)$ are differentiable, then the function defined in Exercise 3.61 is pseudoconvex. (The cases of Exercises 3.62 and 3.63 are similar.) To prove this, we show that for any $x_1, x_2 \in S$, if $\nabla f(x_1)^t(x_2 - x_1) \ge 0$, then $f(x_2) \ge f(x_1)$. From the assumption that $h(x) > 0$, it follows that $\nabla f(x_1)^t(x_2 - x_1) \ge 0$ if and only if $[h(x_1)\nabla g(x_1) - g(x_1)\nabla h(x_1)]^t(x_2 - x_1) \ge 0$. Furthermore, note that $\nabla g(x_1)^t(x_2 - x_1) \le g(x_2) - g(x_1)$, since $g(x)$ is a convex and differentiable function on S, and $\nabla h(x_1)^t(x_2 - x_1) \ge h(x_2) - h(x_1)$, since $h(x)$ is a concave and differentiable function on S. By multiplying the latter inequality by $-g(x_1) \le 0$, and the former one by $h(x_1) > 0$, and adding the resulting inequalities, we obtain (after rearrangement of terms):

$[h(x_1)\nabla g(x_1) - g(x_1)\nabla h(x_1)]^t(x_2 - x_1) \le h(x_1)g(x_2) - g(x_1)h(x_2).$

27

The left-hand side expression is nonegative by our assumption, and therefore, $h(x_1)g(x_2) - g(x_1)h(x_2) \geq 0$, which implies that $f(x_2) \geq f(x_1)$. This completes the proof. $\square$

3.65 For notational convenience let $g(x) = c_1^t x + \alpha_1$, and let $h(x) = c_2^t x + \alpha_2$. In order to prove pseudoconvexity of $f(x) = \dfrac{g(x)}{h(x)}$ on the set $S = \{x : h(x) > 0\}$ we need to show that for any $x_1, \ x_2 \in S$, if $\nabla f(x_1)^t (x_2 - x_1) \geq 0$, then $f(x_2) \geq f(x_1)$.

Assume that $\nabla f(x_1)^t (x_2 - x_1) \geq 0$ for some $x_1, \ x_2 \in S$. By the definition of f, we have $\nabla f(x) = \dfrac{1}{[h(x)]^2}[h(x)c_1 - g(x)c_2]$. Therefore, our assumption yields $[h(x_1)c_1 - g(x_1)c_2]^t (x_2 - x_1) \geq 0$. Furthermore, by adding and subtracting $\alpha_1 h(x_1) + \alpha_2 g(x_1)$ we obtain $g(x_2)h(x_1) - h(x_2)g(x_1) \geq 0$. Finally, by dividing this inequality by $h(x_1)h(x_2)$ (> 0), we obtain $f(x_2) \geq f(x_1)$, which completes the proof of pseudoconvexity of $f(x)$. The psueoconcavity of $f(x)$ on S can be shown in a similar way. Thus, f is pseudolinear. $\square$

CHAPTER 4:

THE FRITZ JOHN AND KARUSH-KUHN-TUCKER
OPTIMALITY CONDITIONS

4.1 $f(x) = xe^{-2x}$. Then $f'(x) = -2xe^{-2x} + e^{-2x} = 0$ implies that $e^{-2x}(1 - 2x) = 0 \Rightarrow x = \bar{x} = 1/2$. Also, $f''(x) = 4e^{-2x}(x - 1)$. Hence, at $\bar{x} = 1/2$, we have that $f''(\bar{x}) < 0$, and so $\bar{x} = 1/2$ is a strict local max for f. This is also a global max and there does not exist a local/global min since from f'', the function is concave for $x \leq 1$ with $f(x) \to -\infty$ as $x \to -\infty$, and f is convex and monotone decreasing for $x \geq 1$ with $f(x) \to 0$ as $x \to \infty$).

4.4 Let $f(x) = 2x_1^2 - x_1 x_2 + x_2^2 - 3x_1 + e^{2x_1 + x_2}$.

a. The first-order necessary condition is $\nabla f(x) = 0$, that is:

$$\begin{bmatrix} 4x_1 - x_2 - 3 + 2e^{2x_1 + x_2} \\ -x_1 + 2x_2 + e^{2x_1 + x_2} \end{bmatrix} = \begin{bmatrix} 0 \\ 0 \end{bmatrix}.$$

The Hessian $H(x)$ of $f(x)$ is

$$H(x) = \begin{bmatrix} 4 + 4e^{2x_1 + x_2} & 2e^{2x_1 + x_2} - 1 \\ 2e^{2x_1 + x_2} - 1 & 2 + e^{2x_1 + x_2} \end{bmatrix},$$ and as can be easily verified,

$H(x)$ is a positive definite matrix for all x. Therefore, the first-order necessary condition is sufficient in this case.

b. $\bar{x} = (0,0)$ is not an optimal solution. $\nabla f(\bar{x}) = [-1 \ 1]^t$, and any direction $d = (d_1, d_2)$ such that $-d_1 + d_2 < 0$ (e.g., $d = (1,0)$) is a descent direction of $f(x)$ at $\bar{x}$.

c. Consider $d = (1,0)$. Then $f(\bar{x} + \lambda d) = 2\lambda^2 - 3\lambda + e^{2\lambda}$. The minimum value of $f(\bar{x} + \lambda d)$ over the interval $[0, \infty)$ is 0.94 and is attained at $\lambda^* = 0.1175$.

d. If the last term is dropped, $f(x) = 2x_1^2 - x_1 x_2 + x_2^2 - 3x_1$. Then the first-order necessary condition yields a unique solution $\bar{x}_1 = 6/7$ and

$\bar{x}_2 = 3/7$. Again, the Hessian of $f(x)$ is positive definite for all x, and so the foregoing values of x_1 and x_2 are optimal. The minimum value of $f(x)$ is given by $-63/49$.

4.5 The KKT system is given by:

$$
\begin{array}{llllllll}
4x_1^3 & +24x_1 & -x_2 & -u_1 & -2u_2 & -u_3 & & = 1 \\
4x_2^3 & +12x_2 & -x_1 & -u_1 & +u_2 & & -u_4 & = 1 \\
x_1 & +x_2 & & & & & & \geq 6 \\
2x_1 & -x_2 & & & & & & \geq 3
\end{array}
$$

$$u_1(6 - x_1 - x_2) = 0, \qquad u_2(3 - 2x_1 + x_2) = 0$$

$$u_3x_1 = 0, \qquad u_4x_2 = 0,$$

$$x_1 \geq 0, \ x_2 \geq 0, \ u_i \geq 0 \text{ for } i = 1, 2, 3, 4.$$

If $\bar{x} = (3, 3)$, then denoting the Lagrange multipliers by $\bar{u}$, we have that $\bar{u}_3 + \bar{u}_4 = 0$. Consequently, the first two equations give $\bar{u}_1 = 152$ and $\bar{u}_2 = 12$. Thus, all the KKT conditions are satisfied at $\bar{x} = (3, 3)$. The Hessian of the objective function is positive definite, and so the problem involves minimizing a strictly convex function over a convex set. Thus, $\bar{x} = (3, 3)$ is the unique global optimum.

4.6 a. In general, the problem seeks a vector y in the column space of A (i.e., $y = Ax$) that is the closest to the given vector b. If b is in the column space of A, then we need to find a solution of the system $Ax = b$. If in addition to this, the rank of A is n, then x is unique. If b is not in the column space of A, then a vector in the column space of A that is the closest to b is the projection of the vector b onto the column space of A. In this case, the problem seeks a solution to the system $Ax = y$, where y is the projection vector of b onto the column space of A. In answers to Parts (b), (c), and (d) below it is assumed that b is not in the column space of A, since otherwise the problem trivially reduces to "find a solution to the system $Ax = b$."

b. Assume that $\|\cdot\|_2$ is used, and let $f(x)$ denote the objective function for this optimization problem. Then, $f(x) = b^t b - 2x^t A^t b + x^t A^t A x$, and the first-order necessary condition is $A^t A x = A^t b$. The Hessian matrix of $f(x)$ is $A^t A$, which is positive semidefinite. Therefore,

30

$f(x)$ is a convex function. By Theorem 4.3.8 it then follows that the necessary condition is also sufficient for optimality.

c. The number of optimal solutions is exactly the same as the number of solutions to the system $A^t Ax = A^t b$.

d. If the rank of A is n, then $A^t A$ is positive definite and thus invertible. In this case, $x = (A^t A)^{-1} A^t b$ is the unique solution. If the rank of A is less than n, then the system $A^t Ax = A^t b$ has infinitely many solutions. In this case, additional criteria can be used to select an appropriate optimal solution as needed. (For details see *Linear Algebra and Its Applications* by Gilbert Strang, Harcourt Brace Jovanovich, Publishers, San Diego, 1988, Third Edition.)

e. The rank of A is 3, therefore, a unique solution exists.
$(A^t A) = \begin{bmatrix} 5 & -2 & 1 \\ -2 & 6 & 4 \\ 1 & 4 & 5 \end{bmatrix}$, and $A^t b = [4 \ 12 \ 12]^t$. The unique solution is $x^* = [2 \ \dfrac{20}{7} \ \dfrac{-2}{7}]^t$.

4.7 a. The KKT system for the given problem is:

$$2x_1 + 2u_1 x_1 + u_2 - u_3 \qquad\qquad = 9/2$$
$$2x_2 - u_1 \quad +u_2 \qquad -u_4 = 4$$
$$x_1^2 - x_2 \le 0$$
$$x_1 + x_2 \le 6$$
$$u_1(x_1^2 - x_2) = 0, \quad u_2(6 - x_1 - x_2) = 0, \quad x_1 u_3 = 0, \quad x_2 u_4 = 0$$
$$x_1 \ge 0, \quad x_2 \ge 0, \quad u_i \ge 0 \text{ for } i = 1, 2, 3, 4.$$

At $\bar{x} = (3/2, 9/4)^t$, denoting the Lagrange multipliers by $\bar{u}$, we necessarily have $\bar{u}_2 = \bar{u}_3 = \bar{u}_4 = 0$, which yields a unique value for u_1 namely, $\bar{u}_1 = 1/2$. The above values for x_1, x_2, and u_i for $i = 1, 2, 3, 4$ satisfy the KKT system, and therefore $\bar{x}$ is a KKT point.

b. Graphical illustration:

From the graph, it follows that at $\bar{x}$, the gradient of $f(x)$ is a negative multiple of the gradient of $g_1(x) = x_1^2 - x_2$, where $g_1(x) \leq 0$ is the only active constraint at $\bar{x}$.

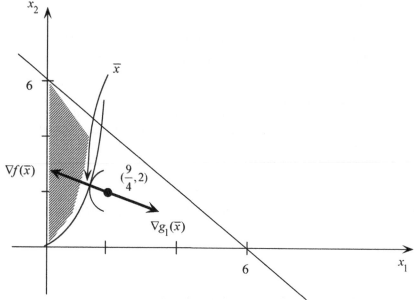

c. It can be easily verified that the objective function is strictly convex, and that the active constraint function is also convex (in fact, the entire feasible region is convex in this case). Hence, $\bar{x}$ is the unique (global) optimal solution to this problem.

4.8 a. The objective function $f(x_1, x_2) = \dfrac{x_1 + 3x_2 + 3}{2x_1 + x_2 + 6}$ is pesudoconvex over the feasible region (see the proof of Lemma 11.4.1). The constraint functions are linear, and are therefore quasiconvex and quasiconcave. Therefore, by Theorem 4.3.8, if $\bar{x}$ is a KKT point for this problem, then $\bar{x}$ is a global optimal solution.

b. First note that $f(0,0) = f(6,0) = 1/2$, and moreover, $f[\lambda(0,0) + (1 - \lambda)(6,0)] = 1/2$ for any $\lambda \in [0,1]$. Since $(0, 0)$ and $(6, 0)$ are feasible solutions, and the feasible region is a polyhedral set, any convex combination of $(0, 0)$ and $(6, 0)$ is also a feasible solution. It is thus sufficient to verify that one of these two points is a KKT point. Consider $(6, 0)$. The KKT system for this problem is as follows:

$$\frac{-5x_2}{(2x_1 + x_2 + 6)^2} + 2u_1 - u_2 - u_3 \qquad = 0$$

$$\frac{5x_1 + 15}{(2x_1 + x_2 + 6)^2} + u_1 + 2u_2 \qquad -u_4 \qquad = 0$$

$$2x_1 + x_2 \leq 12$$

$$-x_1 + 2x_2 \leq 4$$

$$x_1 \geq 0, \ x_2 \geq 0, \ u_i \geq 0 \text{ for } i = 1, 2, 3, 4$$

$$u_1(2x_1 + x_2 - 12) = 0, \ u_2(-x_1 + 2x_2 - 4) = 0, \ u_3 x_1 = 0, \ u_4 x_2 = 0.$$

Substituting $(6, 0)$ for (x_1, x_2) into this KKT system yields the following unqiue values for the Lagrangian multipliers: $u_1 = u_2 = u_3 = 0$, and $u_4 = 5/36$. Since $u_i \geq 0, \ \forall i = 1,2,3,4$, we conclude that $(6, 0)$ is indeed a KKT point, and therefore, by Part (a), it solves the given problem. Hence, by the above argument, any point on the line segment joining $(0, 0)$ and $(6, 0)$ is an optimal solution.

4.9 Note that $c \neq 0$, as given.

a. Let $f(d) = -c^t d$, and $g(d) = d^t d - 1$. The KKT system for the given problem is as follows:

$$-c + 2du = 0$$

$$d^t d \leq 1$$

$$u(d^t d - 1) = 0$$

$$u \geq 0.$$

$d = \bar{d} \equiv c/\|c\|$ and $u = \bar{u} \equiv \|c\|/2$ yields a solution to this system. Hence, $\bar{d}$ is a KKT point. Moreover, $\bar{d}$ is an optimal solution, because it is a KKT point and sufficiency conditions for optimality are met since $f(d)$ is a linear function, hence it is pseudoconvex, and $g(d)$ is a convex function, hence it is quasiconvex. Furthermore, $\bar{d}$ is the unique global optimal solution since the KKT system provides necessary and sufficient conditions for optimality in this case, and $\bar{d} = c/\|c\|$, $\bar{u} = \|c\|/2$ is its unique solution. To support this statement, notice that if $u > 0$, then $d^t d = 1$, which together with the first equation results in $d = c/\|c\|$ and $u = \|c\|/2$. If $u = 0$, then the first equation is inconsistent regardless of d since $c \neq 0$.

33

b.   The steepest ascent direction of a differentiable function $f(x)$ at $\bar{x}$ can be found as an optimal solution $\bar{d}$ to the following problem:

$$\text{Maximize } \{\nabla f(\bar{x})^t d : d^t d \le 1\},$$

which is identical to the problem considered in Part (a) with $c = \nabla f(\bar{x})$. Thus, if $\nabla f(\bar{x}) \ne 0$, then the steepest ascent direction is given by $\bar{d} = \nabla f(\bar{x})/\|\nabla f(\bar{x})\|$.

4.10 a.   In order to determine whether a feasible solution $\bar{x}$ is a KKT point, one needs to examine if there exists a feasible solution to the system:

$$\nabla f(\bar{x}) + \sum_{i \in I} \nabla g_i(\bar{x})u_i = 0, \ u_i \ge 0 \text{ for } i \in I,$$

where I is the set of indices of constraints that are active at $\bar{x}$.

Let $c = -\nabla f(\bar{x})$ and let $A^t = [\nabla g_i(\bar{x}), i \in I]$. Then the KKT system can be rewritten as follows:

$$A^t u = c, \ u \ge 0. \tag{1}$$

Therefore, $\bar{x}$ is a KKT point if and only if System (1) has a solution. Note that System (1) is linear, and it has a solution if and only if the optimal objective value in the following problem is zero:

Minimize $e^t y$

subject to $A^t u \pm y = c$

$\qquad\qquad u \ge 0, \ y \ge 0,$

where e is a column vector of ones, y is a vector of artificial variables, and where $\pm y$ denotes that the components of y are ascribed the same sign as that of the respective components of c. This problem is a Phase I LP for finding a nonnegative solution to $A^t u = c$.

b.   In the presence of equality constraints $h_i(x) = 0$, $i = 1,...,\ell$, the KKT system is given by

$$\nabla f(\bar{x}) + \sum_{i \in I} \nabla g_i(\bar{x})u_i + \sum_{i=1}^{\ell} \nabla h_i(\bar{x})v_i = 0, \ u_i \ge 0 \text{ for } i \in I,$$

where I is the set of indices of the inequality constraints that are active at $\bar{x}$. Let A^t be as defined in Part (a), and let $B^t = [\nabla h_i(\bar{x}), \; i = 1,...,\ell]$ be the $n \times \ell$ Jacobian matrix at $\bar{x}$ for the equality constraints. Then the corresponding Phase I problem is given as follows:

Minimize $\quad e^t y$

subject to $\quad A^t u + B^t v \pm y = c$

$\qquad\qquad u \geq 0, \; y \geq 0.$

c. In this example, we have $\bar{x} = (1,2,5)^t$, $c = -\nabla f(\bar{x}) = -(8,3,23)^t$. Furthermore, $I = \{1,3\}$, and therefore,

$$A^t = [\nabla g_1(\bar{x}) \; \nabla g_3(\bar{x})] = \begin{bmatrix} 2 & -1 \\ 4 & -1 \\ -1 & 0 \end{bmatrix}. \text{ Thus, } \bar{x} = (1,2,5)^t \text{ is a KKT}$$

point if and only if the optimal objective value of the following problem is zero:

Minimize $\quad y_1 \;\; + \;\; y_2 \; + y_3$

subject to $\quad -2u_1 \; + \;\; u_3 \; + y_1 \qquad\qquad\qquad = 8$

$\qquad\qquad -4u_1 \; + \;\; u_3 \qquad\quad +y_2 \qquad\qquad = 3$

$\qquad\qquad \;\; u_1 \qquad\qquad\qquad\qquad\quad +y_3 \;\; = 23$

$\qquad\qquad u_1 \geq 0, \; u_3 \geq 0, \; y_i \geq 0 \text{ for } i = 1, 2, 3.$

However, the optimal solution to this problem is given by $\bar{u}_1 = 2.5$, $\bar{u}_3 = 13$, $\bar{y}_1 = \bar{y}_2 = 0$, $\bar{y}_3 = 20.5$, and the optimal objective value is positive (20.5), and so we conclude that $\bar{x} = (1,2,5)^t$ is not a KKT point.

4.12 Let $\quad y_j = \dfrac{a_j}{b} x_j$ and $d_j = \dfrac{c_j a_j}{b}$ for $j = 1,..., n$. Then the given optimization problem is equivalent to the following, re-written in a more convenient form:

Minimize $\quad \displaystyle\sum_{j=1}^{n} \left(\dfrac{d_j}{y_j} \right)$

subject to $\displaystyle\sum_{j=1}^{n} y_j = 1$

$$y_j \geq 0 \text{ for } j = 1,\ldots, n.$$

The KKT system for the above problem is given as follows:

$$\frac{-d_j}{y_j^2} + v - u_j = 0 \text{ for } j = 1,\ldots, n$$

$$u_j y_j = 0, \ u_j \geq 0, \text{ and } y_j \geq 0 \text{ for } j = 1,\ldots, n.$$

Readily, for each $j = 1,\ldots, n$, y_j must take on a positive value, and hence $u_j = 0, \ \forall j = 1,\ldots, n$. The KKT system thus yields $y_j = \dfrac{\sqrt{d_j}}{\sqrt{v}}$, $\forall j$, which upon summing and using $\displaystyle\sum_{j=1}^{n} y_j = 1$ gives $v = \left[\displaystyle\sum_{j=1}^{n} \sqrt{d_j}\right]^2$. Thus $(\bar{y}, \bar{v}, \bar{u})$ given by $\bar{y}_j = \dfrac{\sqrt{d_j}}{\displaystyle\sum_{j=1}^{n}\sqrt{d_j}}$, $\forall j = 1,\ldots, n$, $\bar{v} = \left[\displaystyle\sum_{j=1}^{n}\sqrt{d_j}\right]^2$, and $\bar{u}_j = 0, \ \forall j = 1,\ldots, n$, is the (unique) solution to the above KKT system. The unique KKT point for the original problem is thus given by

$$\bar{x}_j = \frac{b\sqrt{a_j c_j}}{a_j \displaystyle\sum_{j=1}^{n}\sqrt{a_j c_j}}, \ \forall j = 1,\ldots, n.$$

4.15 Consider the problem

Minimize $\displaystyle\sum_{j=1}^{n} x_j$

subject to $\displaystyle\prod_{j=1}^{n} x_j = b,$

$$x_j \geq 0, \ \forall j = 1,\ldots, n,$$

where b is a positive constant. Since feasibility requires $x_j > 0$, $\forall j = 1,...,n$, the only active constraint is the equality restriction, and because of the linear independence constraint qualification, the KKT conditions are necessary for optimality. The KKT system for this problem is thus given as follows:

$$1 + v \prod_{\substack{i=1 \\ i \neq j}}^{n} x_i = 0 \text{ for } j = 1,..., n$$

$$\prod_{j=1}^{n} x_j = b.$$

By multiplying the jth equation by x_j for $j = 1,...,n$, and noting that $\prod_{j=1}^{n} x_j = b$, we obtain

$$x_j + vb = 0 \text{ for } j = 1,..., n.$$

Therefore, $\sum_{j=1}^{n} x_j + nbv = 0$, which gives the unique value for the Lagrange multiplier $v = -\sum_{j=1}^{n} x_j / nb$. By substituting this expression for v into each of the equations $x_j + vb = 0$ for $j = 1,..., n$, we then obtain

$$x_j = \frac{1}{n} \sum_{k=1}^{n} x_k \text{ for } j = 1,..., n.$$ This necessarily implies that the values of x_j are all identical, and since $\prod_{j=1}^{n} x_j = b$, we have that $\bar{x}_j = b^{1/n}$, $\forall j = 1,...,n$ yields the unique KKT solution, and since the KKT conditions are necessary for optimality, this gives the unique optimum to the above problem. Therefore, $\frac{1}{n} \sum_{j=1}^{n} \bar{x}_j = b^{1/n}$ is the optimal objective function value. We have thus shown that for any positive vector x such that $\prod_{j=1}^{n} x_j = b$, we have that

$$\frac{1}{n} \sum_{j=1}^{n} x_j \geq \text{minimum} \left\{ \frac{1}{n} \sum_{j=1}^{n} x_j : \prod_{j=1}^{n} x_j = b \right\} = b^{1/n} = \left(\prod_{j=1}^{n} x_j \right)^{1/n}.$$

But, for any given positive vector x, the product of its components is a constant, and so the above inequality implies that $\frac{1}{n} \sum_{j=1}^{n} x_j \geq \left(\prod_{j=1}^{n} x_j \right)^{1/n}$.

Furthermore, if any component is zero, then this latter inequality holds trivially. $\square$

4.27 a. $d = 0$ is a feasible solution and it gives the objective function value equal to 0. Therefore, $\bar{z} \leq 0$.

b. If $\bar{z} < 0$, then $\nabla f(\bar{x})^t \bar{d} < 0$. By Theorem 4.1.2, $\bar{d}$ is a descent direction. Furthermore, by the concavity of $g_i(x)$ at $\bar{x}$, $i \in I$, since $g_i(\bar{x}) = 0$, there exists a $\delta > 0$ such that $g_i(\bar{x} + \lambda \bar{d}) \leq \lambda \nabla g_i(\bar{x})^t \bar{d}$ for $\lambda \in (0, \delta)$. Since the vector $\bar{d}$ is a feasible solution to the given problem, we necessarily have $\nabla g_i(\bar{x})^t \bar{d} \leq 0$ for $i \in I$, and thus $g_i(\bar{x} + \lambda \bar{d}) \leq 0$ for $\lambda \in (0, \delta)$. All the remaining constraint functions are continuous at $\bar{x}$, and so again there exists a $\delta_1 > 0$ such that $g_i(\bar{x} + \lambda \bar{d}) \leq 0$ for $\lambda \in (0, \delta_1)$, $\forall i = 1, ..., m$. This shows that $\bar{d}$ is a feasible descent direction at $\bar{x}$.

c. If $\bar{z} = 0$, then the dual to the given linear program has an optimal solution of objective function value zero. This dual problem can be formulated as follows:

Maximize $\quad -v_1^t e - v_2^t e$

subject to $\quad -\sum_{i \in I} \nabla g_i(\bar{x}) u_i - v_1 + v_2 = \nabla f(\bar{x})$

$\quad u_i \geq 0 \text{ for } i \in I, \ v_1 \geq 0, \ v_2 \geq 0,$

where $e \in R^n$ is a vector of ones. Thus if $\bar{z} = 0$, then v_1 and v_2 are necessarily equal to 0 at an optimal dual solution, and so there exist nonnegative numbers u_i, $i \in I$, such that $\nabla f(\bar{x}) + \sum_{i \in I} \nabla g_i(\bar{x}) u_i = 0$.

Thus, $\bar{x}$ satisfies the KKT conditions.

4.28 Consider the unit simplex $S = \{y : e^t y = 1, \ y \geq 0\}$, which is essentially an $(n - 1)$-dimensional body. Its center is given by $y_0 = (\frac{1}{n}, ..., \frac{1}{n})^t$.

Examine a (maximal) sphere with center y_0 and radius r that is inscribed with S. Then, r is the distance from y_0 to the center of the one less dimensional simplex, say, formed in the $(y_1, \ldots, y_{n-1})$-space, where the latter center in the full y-space is thus given by $(\frac{1}{n-1}, \ldots, \frac{1}{n-1}, 0)$. Hence, we get

$$r^2 = (n-1)\left[\frac{1}{(n-1)} - \frac{1}{n}\right]^2 + \frac{1}{n^2} = \frac{1}{n(n-1)}.$$

Therefore, the given problem examines the $(n-1)$-dimensional sphere formed by the intersection of the sphere given by $\|y - y_0\|^2 \le r^2$ with the hyperplane $e^t y = 1$, without the nonnegativity restrictions $y \ge 0$, and seeks the minimal value of any coordinate in this region, say, that of y_1. The KKT conditions for this problem are as follows:

$$2(y_1 - y_{01})u_0 + v = -1$$
$$2(y_i - y_{0i})u_0 + v = 0 \quad \text{for } i = 2, \ldots, n$$
$$\|y - y_0\|^2 \le 1/n(n-1)$$
$$e^t y = 1$$
$$u_0 \ge 0, \ u_0\left(\|y - y_0\|^2 - \frac{1}{n(n-1)}\right) = 0.$$

Let $\bar{y} = [0, \frac{1}{n-1}, \ldots, \frac{1}{n-1}]^t$. To show that $\bar{y}$ is a KKT point for this problem, all that one needs to do is to substitute $\bar{y}$ for y in the foregoing KKT system and verify that the resulting system in (u_0, v) has a solution.

Readily, $e^t \bar{y} = \frac{n-1}{n-1} = 1$, and $\bar{y} - y_0$ has $(n-1)$ coordinates equal to $\frac{1}{n(n-1)}$, and one coordinate (the first one) equal to $-\frac{1}{n}$, so that $\|\bar{y} - y_0\|^2 = \frac{1}{n(n-1)}$. This means that $\bar{y}$ is a feasible solution. Moreover, the equations for indices 2 through n of the KKT system yield

$v = -\dfrac{2u_0}{n(n-1)}$, which together with the first equation gives

$u_0 = \dfrac{n-1}{2} \geq 0$. Thus, $\bar{y}$ is a KKT point for this problem. Since the problem is a convex program, this is an optimal solution. Thus, since this is true for minimizing any coordinate of y, even without the nonnegativity constraints present explicitly, the intersection is embedded in the nonnegative orthant.

4.30 Substitute $y = x - \bar{x}$ to obtain the following equivalent form of Problem $\overline{P}$:

$$\text{Minimize } \left\{ \|y - d\|^2 : Ay = 0 \right\}.$$

a. Problem $\overline{P}$ seeks a vector in the nullspace of A that is closest to the given vector d, i.e., to the vector $-\nabla f(\bar{x})$. Since the rank of A is m, an optimal solution to the problem $\overline{P}$ is the orthogonal projection of the vector $-\nabla f(\bar{x})$ onto the nullspace of A (i.e., start from $\bar{x}$, take a unit step along $-\nabla f(\bar{x})$, and then project the resulting point orthogonally back onto the constraint surface $Ax = b$).

b. The KKT conditions for Problem $\overline{P}$ are as follows:

$$\begin{aligned} x + A^t v &= \bar{x} + d \\ Ax &= b. \end{aligned}$$

The objective function of $\overline{P}$ is strictly convex, and the constraints are linear, and so the KKT conditions for Problem $\overline{P}$ are both necessary and sufficient for optimality.

c. If $\bar{x}$ is a KKT point for Problem $\overline{P}$, then there exists a vector $\bar{v}$ of Lagrange multipliers associated with the equations $Ax = b$, such that

$$A^t \bar{v} = d, \text{ that is, } \nabla f(\bar{x}) + A^t \bar{v} = 0.$$

Hence, $\bar{x}$ is a KKT point for Problem P provided $\bar{v} \geq 0$.

d. From the KKT system, we get

$$\hat{x} = \bar{x} + d - A^t v \tag{1}$$

Multiplying (1) by A and using $A\hat{x} = A\bar{x} = b$, we get

40

$$AA^t v = Ad.$$

Since A is of full row rank, the $(m \times m)$ matrix AA^t is nonsingular. Thus, $v = (AA^t)^{-1} Ad$. Substituting this into (1), we get $\hat{x} = \bar{x} + d - A^t (AA^t)^{-1} Ad$.

4.31 Let $c^t = \nabla_N f(\bar{x})^t - \nabla_B f(\bar{x})^t B^{-1} N$. The considered direction finding problem is a linear program in which the function $c^t d$ is to be minimized over the region $\{d : 0 \le d_j \le 1, \ j \in J\}$, where J is the set of indices for the nonbasic variables. It is easy to verify that $c^t \bar{d}_N \le 0$ at optimality. In fact, an optimal solution to this problem is given by: $\bar{d}_j = 0$ if $c_j \ge 0$, and $\bar{d}_j = 1$ if $c_j < 0$, $\forall j \in J$. To verify if $\bar{d}$ is an improving direction, we need to examine if $\nabla f(\bar{x})^t \bar{d} < 0$, where

$$\nabla f(\bar{x})^t \bar{d} = [\nabla_B f(\bar{x})^t \ \nabla_N f(\bar{x})^t] \begin{bmatrix} \bar{d}_B \\ \bar{d}_N \end{bmatrix} =$$
$$[-\nabla_B f(\bar{x})^t B^{-1} N + \nabla_N f(\bar{x})^t] \bar{d}_N = c^t \bar{d}_N.$$

Therefore, $c^t \bar{d}_N < 0$ implies that $\nabla f(\bar{x})^t \bar{d} < 0$. Hence, if $\bar{d} \neq 0$, then we must have $\bar{d}_N \neq 0$ (else $\bar{d}_B = 0$ as well), whence $c^t \bar{d}_N < 0$ from above. This means that $\bar{d}$ is an improving direction at $\bar{x}$. Moreover, to show that $\bar{d}$ is a feasible direction at $\bar{x}$, first, note that

$$A\bar{d} = [B \ N] \begin{bmatrix} -B^{-1} N \bar{d}_N \\ \bar{d}_N \end{bmatrix} = -N\bar{d}_N + N\bar{d}_N = 0, \qquad \text{and} \qquad \text{therefore,}$$

$A(\bar{x} + \lambda \bar{d}) = b$ for all $\lambda \ge 0$. Moreover,

$$\bar{x} + \lambda \bar{d} = \begin{bmatrix} B^{-1} b - \lambda B^{-1} N \bar{d}_N \\ \lambda \bar{d}_N \end{bmatrix} \ge 0$$

for $\lambda > 0$ and sufficiently small since $B^{-1} b > 0$ and $\bar{d}_N \ge 0$, which implies that $\bar{x} + \lambda \bar{d} \ge 0$ for all $0 \le \lambda \le \bar{\lambda}$, where $\bar{\lambda} > 0$. Thus, $\bar{d}$ is a

41

feasible direction at $\bar{x}$. Hence, $\bar{d} \neq 0$ implies that $\bar{d}$ is an improving feasible direction.

Finally, suppose that $\bar{d} = 0$, which means that $\bar{d}_N = 0$. Then $c \geq 0$. The KKT conditions at $\bar{x}$ for the original problem can then be written as follows:

$$\nabla_B f(\bar{x}) - u_B + B^t v = 0$$
$$\nabla_N f(\bar{x}) - u_N + N^t v = 0$$
$$u_B^t \bar{x}_B = 0, \ u_N^t \bar{x}_N = 0, \ u_B \geq 0, \ u_N \geq 0.$$

Let $\bar{u}_B = 0$, $\bar{v}^t = -\nabla_B f(\bar{x})^t B^{-1}$, and $\bar{u}_N^t = \nabla_N f(\bar{x})^t - \nabla_B f(\bar{x})^t B^{-1} N$. Simple algebra shows that $(\bar{u}_B, \bar{u}_N, \bar{v})$ satisfies the above system (solve for v from the first equation and substitute it in the second equation). Therefore, $\bar{x}$ is a KKT point whenever $\bar{d} = 0$ (and is optimal if, for example, f is pseudoconvex). $\square$

4.33 In the first problem, the KKT system is given by:

$$c + Hx + A^t u = 0 \tag{1}$$
$$Ax + y = b \tag{2}$$
$$u^t y = 0$$
$$x \geq 0, \ y \geq 0, \ u \geq 0.$$

Since the matrix H is invertible, Equation (1) yields $H^{-1}c + x + H^{-1}A^t u = 0$. By premultiplying this equation by A, we obtain $AH^{-1}c + Ax + AH^{-1}A^t u = 0$, which can be rewritten as

$$AH^{-1}c + b + Ax - b + AH^{-1}A^t u = 0. \tag{3}$$

Next, note that from Equation (2), we have $y = b - Ax$, so that Equation (3) can be further rewritten as

$$h + Gu - y = 0, \text{ where } u^t y = 0, \ u \geq 0, \ y \geq 0. \tag{4}$$

In the second given problem, the KKT system is given by

$$h + Gv - z = 0, \ v^t z = 0, \ v \geq 0, \ z \geq 0, \tag{5}$$

42

where z is the vector of Lagrange multipliers. By comparing (4) and (5), we see that the two problems essentially have identical KKT systems, where $u \equiv v$ and $y \equiv z$, that is, the Lagrange multipliers in the first problem are decision variables in the second problem, while the Lagrange multipliers in the second problem are slack variables in the first problem.

4.37 We switch to minimizing the function $f(x_1, x_2) = -x_1^2 - 4x_1 x_2 - x_2^2$.

a. The KKT system is as follows:

$$
\begin{aligned}
-2x_1 \quad -4x_2 \quad +2vx_1 &= 0 \\
-4x_1 \quad -2x_2 \quad +2vx_2 &= 0 \\
x_1^2 \quad +x_2^2 \quad &= 1.
\end{aligned}
$$

There are four solutions to this system:

$$
\begin{aligned}
(x_1, x_2) &= (1/\sqrt{2}, 1/\sqrt{2}), \text{ and } v = 3 \\
(x_1, x_2) &= (-1/\sqrt{2}, -1/\sqrt{2}), \text{ and } v = 3 \\
(x_1, x_2) &= (1/\sqrt{2}, -1/\sqrt{2}), \text{ and } v = -1 \\
(x_1, x_2) &= (-1/\sqrt{2}, 1/\sqrt{2}), \text{ and } v = -1.
\end{aligned}
$$

The objective function $f(x_1, x_2)$ takes on the value of -3 for the first two points, and the value of 1 at the remaining two. Since the linear independence constraint qualification (CQ) holds, the KKT conditions are necessary for optimality. Hence, there are two optimal solutions: $\bar{x}_1 = (1/\sqrt{2}, 1/\sqrt{2})$ and $\bar{x}_2 = (-1/\sqrt{2}, -1/\sqrt{2})$. To support this statement, one can use a graphical display, or use the second-order sufficiency condition given in Part (b) below.

b. $L(x) = -x_1^2 - 4x_1 x_2 + v(x_1^2 + x_2^2 - 1)$. Therefore,

$$
\nabla^2 L(x) = 2 \begin{bmatrix} v - 1 & -2 \\ -2 & v - 1 \end{bmatrix}.
$$

For $v = 3$, $\nabla^2 L(x)$ is a positive definite matrix and therefore, $\bar{x}_1 = (\sqrt{2}/2, \sqrt{2}/2)$ and $\bar{x}_2 = (-\sqrt{2}/2, -\sqrt{2}/2)$ are both strict local optima.

c.   See answers to Parts (a) and (b).

4.41 a.   See the proof of Lemma 10.5.3.

b.   See the proof of Theorem 10.5.4.

c.   Let P_d denote the given problem and note that since this problem is convex, the KKT conditions are sufficient for optimality. Hence, it is sufficient to produce a KKT solution to Problem P_d of the given form $\hat{d}$. Toward this end, consider the KKT conditions for Problem P_d:

$$\nabla f(\overline{x}) + A_1^t v + 2du = 0 \tag{1}$$
$$A_1 d = 0 \tag{2}$$
$$\|d\|^2 \leq 1, \; u \geq 0, \; u(\|d\|^2 - 1) = 0.$$

Premultiplying (1) by A_1 and using (2), we get

$$A_1 \nabla f(\overline{x}) + A_1 A_1^t v = 0.$$

Since A_1 is of full (row) rank, $A_1 A_1^t$ is nonsingular, and so we get

$$v = -(A_1 A_1^t)^{-1} A_1 \nabla f(\overline{x}). \tag{3}$$

Thus, (1) yields

$$2du = -P\nabla f(\overline{x}) = \overline{d}. \tag{4}$$

Hence, if $\overline{d} = 0$, we can take $d = \hat{d} = 0$ and $u = 0$, which together with (3) yields $\hat{d} = 0$ as a KKT point (hence, an optimum to P_d, with say, $\lambda \equiv 1$). On the other hand, if $\overline{d} \neq 0$, then let $\hat{d} = \dfrac{\overline{d}}{\|\overline{d}\|}$, $u = \dfrac{\|\overline{d}\|}{2}$, and let v be given by (3). Thus, noting that $A_1 \hat{d} = 0$ since $A_1 P = 0$, we get that $\hat{d}$ is a KKT point and hence an optimum to P_d (with $\lambda \equiv \|\overline{d}\| > 0$).

d. If $A = -I_n$, then A_1 is an $m \times n$ submatrix of $-I_n$, where m is the number of variables that are equal to zero at the current solution $\bar{x}$.

Then $\quad A_1 A_1^t = I_m, \quad$ and $\quad A_1^t A_1 = \begin{bmatrix} I_m & 0 \\ 0 & 0 \end{bmatrix}.$ $\quad$ Therefore,

$P = \begin{bmatrix} 0 & 0 \\ 0 & I_{n-m} \end{bmatrix}$, and $\bar{d}_j = 0$ if $\bar{x}_j = 0$, and $\bar{d}_j = -\dfrac{\partial f(\bar{x})}{\partial x_j}$ if

$\bar{x}_j > 0$. Hence, $\bar{d}$ is the projection of $-\nabla f(\bar{x})$ onto the nullspace of the active (nonnegativity) constraints.

4.43 Note that $C = \{d : Ad = 0\}$ is the nullspace of A, and P is the projection matrix onto the nullspace of A. If $d \in C$, then $Pd = d$, and so, $d = Pw$ with $w \equiv d$. On the other hand, if $d = Pw$ for some $w \in R^n$, we have that $Ad = APw = 0$ since $AP = 0$. Hence, $d \in C$. This shows that $d \in C$ if and only if there exists a $w \in R^n$ such that $Pw = d$. Next, we show that if H is a symmetric matrix, then $d^t Hd \geq 0$ for all $d \in C$ if and only if $P^t HP$ is positive semidefinite.

$(\Rightarrow)$ Suppose that $d^t Hd \geq 0$ for all $d \in C$. Consider any $w \in R^n$ and let $d = Pw$. Then $Ad = APw = 0$ since $AP = 0$, and so $d \in C$. Thus $d^t Hd \geq 0$, which yields $w^t P^t HPw \geq 0$ for any $w \in R^n$. Hence, the matrix $P^t HP$ is positive semidefinite.

$(\Rightarrow)$ If $w^t P^t HPw \geq 0$ for all $w \in R^n$, then in particular for any $d \in C$, we have $d^t P^t HPd \geq 0$, which gives $d^t Hd \geq 0$ since for any $d \in C$ we have $Pd = d$. $\quad \square$

CHAPTER 5:

CONSTRAINT QUALIFICATIONS

5.1 Let T denote the cone of tangents of S at $\bar{x}$ as given in Definition 5.1.1.

a. Let W denote the set of directions defined in this part of the exercise. That is, $d \in W$ if there exists a nonzero sequence $\{\beta_k\}$ convergent to zero, and a function $\alpha:R \to R^n$ that converges to 0 as $\beta \to 0$, such that $\bar{x} + \beta_k d + \beta_k \alpha(\beta_k) \in S$ for any k. We need to show that $W = T$. First, note that $0 \in W$ and $0 \in T$. Now, let d be a nonzero vector from the set T. Then there exist a positive sequence $\{\lambda_k\}$ and a sequence $\{x_k\}$ of points from S convergent to $\bar{x}$ such that $d = \lim_{k \to \infty} \lambda_k(x_k - \bar{x})$. Without loss of generality, assume that $x_k \neq \bar{x}, \forall k$ (since $d \neq 0$). Therefore, for this sequence $\{x_k\}$, consider the nonzero sequence $\{\beta_k\}$ such that $\beta_k d$ is the projection of $x_k - \bar{x}$ onto the vector d. Hence, $\{\beta_k\} \to 0^+$. Furthermore, let $y_k \equiv (x_k - \bar{x}) - \beta_k d$. Because of the projection operation, we have that

$$\left\| x_k - \bar{x} \right\|^2 = \beta_k^2 \left\| d \right\|^2 + \left\| y_k \right\|^2,$$

i.e.,
$$\frac{\left\| y_k \right\|^2}{\beta_k^2} = \left\| d \right\|^2 \left[\frac{\left\| x^k - \bar{x} \right\|^2}{\beta_k^2 \left\| d \right\|^2} - 1 \right]. \tag{1}$$

But we have that $\dfrac{\beta_k \left\| d \right\|}{\left\| x^k - \bar{x} \right\|} = \cos(\gamma_k)$, where γ_k is the angle between $(x_k - \bar{x})$ and d. Since $d \in T$, we have that $\gamma_k \to 0$ and so $\cos(\gamma_k) \to 1$ and thus $\dfrac{\left\| y_k \right\|}{\beta_k} \to 0$ from (1). Consequently, we can define $\alpha:R \to R^n$ such that $\beta_k \alpha(\beta_k) = y_k$ so that $x_k = \bar{x} + \beta_k d +$

$\beta_k \alpha(\beta_k) \in S$, $\forall k$, with $\alpha(\beta_k) = y_k / \beta_k \to 0$ as $\beta_k \to 0$. Hence, $d \in W$.

Next, we show that if $d \in W$, then $d \in T$. For this purpose, let us note that if $d \in W$, then the sequence $\{x_k\} \subseteq S$, where $x_k = \bar{x} + \beta_k d + \beta_k \alpha(\beta_k)$, converges to $\bar{x}$, and moreover, the sequence $\left\{ \dfrac{1}{\beta_k}(x_k - \bar{x}) - d \right\}$ converges to the zero vector. This shows that there exists a sequence $\{\lambda_k\}$, where $\lambda_k = \dfrac{1}{\beta_k}$, and a sequence $\{x_k\}$ of points from S convergent to $\bar{x}$ such that $d = \lim\limits_{k \to \infty} \lambda_k (x_k - \bar{x})$. This means that $d \in T$, and so the proof is complete.

b. Again, let W denote the set of directions defined in this part of the exercise. That is, $d \in W$ if there exists a nonnegative scalar λ and a sequence $\{x_k\}$ of points from S convergent to $\bar{x}$, $x_k \neq \bar{x}$ for all k, such that $d = \lim\limits_{k \to \infty} \lambda \dfrac{x_k - \bar{x}}{\|x_k - \bar{x}\|}$. Again in this case, we have $0 \in W$ and $0 \in T$, and so let d be a nonzero vector in T. Then there exists a sequence $\{x_k\}$ of points from S different from $\bar{x}$ and a positive sequence $\{\lambda_k\}$ such that $x_k \to \bar{x}$, and $d = \lim\limits_{k \to \infty} \lambda_k \|x_k - \bar{x}\| \dfrac{x_k - \bar{x}}{\|x_k - \bar{x}\|}$. Under the assumption that $d \in T$, the sequence $\left\{ \lambda_k \|x_k - \bar{x}\| \right\}$ is contained in a compact set. Therefore, it must have a convergent subsequence. Without loss of generality, assume that the sequence $\left\{ \dfrac{x_k - \bar{x}}{\|x_k - \bar{x}\|} \right\}$ itself is convergent. If so, then we conclude that $d = \lim\limits_{k \to \infty} \lambda \dfrac{x_k - \bar{x}}{\|x_k - \bar{x}\|}$, where $\lambda = \|d\|$. Hence, $d \in W$.

47

Conversely, let $d \in W$, where again, $d \neq 0$. Then we can simply take

$$\lambda_k = \frac{\lambda}{\left\| x_k - \bar{x} \right\|} > 0$$ to readily verify that $d \in T$. This completes the

proof. $\square$

5.12 a. See the proof of Theorem 10.1.7.

b. By Part (a), $\bar{x}$ is a FJ point. Therefore, there exist scalars u_0 and u_i for $i \in I$, such that

$$u_0 \nabla f(\bar{x}) + \sum_{i \in I} u_i \nabla g_i(\bar{x}) = 0,$$

$$u_0 \geq 0, \ u_i \geq 0 \ \text{for} \ i \in I, \ (u_0, u_i \ \text{for} \ i \in I) \neq 0.$$

If $u_0 = 0$, then the system

$$\sum_{i \in I} u_i \nabla g_i(\bar{x}) = 0,$$

$$u_i \geq 0 \ \text{for} \ i \in I$$

has a nonzero solution. Then, by Gordan's Theorem, no vector d exists such that $\nabla g_i(\bar{x})^t d < 0$ for all $i \in I$. This means that $G_0 = \varnothing$, and so $c\ell(G_0) = \varnothing$, whereas $G' \neq \varnothing$ (since $0 \in G'$). This contradicts Cottle's constraint qualification.

5.13 a. $\bar{x} = [1 \ 0]^t$, $I = \{1, 2\}$, $\nabla g_1(\bar{x}) = [2 \ 0]^t$, $\nabla g_2(\bar{x}) = [0 \ -1]^t$. The gradients of the binding constraints are linearly independent; hence, the linear independence constraint qualification holds. This implies that Kuhn-Tucker's constraint qualification also holds (see Figure 5.2 in the text and its associated comments).

b. If $\bar{x} = [1 \ 0]^t$, then the KKT conditions yields:

$$\begin{aligned} -1 \quad + 2u_1 \quad &= 0 \\ - u_2 &= 0, \end{aligned}$$

i.e., $u_1 = \dfrac{1}{2}$ and $u_2 = 0$. Since the Lagrange multipliers are nonnegative, we conclude that $\bar{x}$ is a KKT point.

48

Note that a feasible solution must be in the unit circle centered at the origin; hence no feasible solution can have its first coordinate greater than 1. Therefore, $\bar{x}$ (which yields the objective value of -1) is the global optimal solution.

5.15 X is an open set, the functions of nonbinding constraints are continuous at $\bar{x}$, and the functions whose indices are in the set J are pseudoconcave at $\bar{x}$. Therefore, by the same arguments as those used in the proof of Lemma 4.2.4, any vector d that satisfies the inequalities $\nabla g_i(\bar{x})^t d \leq 0$ for $i \in J$, and $\nabla g_i(\bar{x})^t d < 0$ for $i \in I - J$ is a feasible direction at $\bar{x}$. Hence, if $\bar{x}$ is a local minimum, then the following system has no solution:

$$\nabla f(\bar{x})^t d < 0$$
$$\nabla g_i(\bar{x})^t d < 0 \text{ for } i \in I - J$$
$$\nabla g_i(\bar{x})^t d \leq 0 \text{ for } i \in J.$$

Accordingly, consider the following pair of primal and dual programs P and D, where $y_0 \in R$ is a dummy variable:

P: Maximize y_0

subject to $\nabla f(\bar{x})^t d + y_0 \leq 0$

$\nabla g_i(\bar{x})^t d + y_0 \leq 0, \quad \forall i \in I - J$

$\nabla g_i(\bar{x})^t d \leq 0, \quad \forall i \in J.$

D: Minimize 0

subject to $u_0 \nabla f(\bar{x}) + \sum_{i \in I} u_i \nabla g_i(\bar{x}) = 0$ \hfill (1)

$u_0 + \sum_{i \in I - J} u_i = 1$ \hfill (2)

$(u_0, u_i \text{ for } i \in I) \geq 0.$ \hfill (3)

Then, since the foregoing system has no solution, then we must have that P has an optimal value of zero (since if $y_0 > 0$ for a feasible solution (y_0, d), then P is unbounded), which means that D is feasible, i.e., (1) – (3) has a solution. If $u_0 > 0$ in any such solution, then $\bar{x}$ is a KKT point and we are done. Else, suppose that $u_0 = 0$, which implies by (2) that $I - J \neq \varnothing$. Furthermore, letting d belong to the given nonempty set in the exercise such that $\nabla g_i(\bar{x})^t d \leq 0$ for $i \in J$, and $\nabla g_i(\bar{x})^t d < 0$ for $i \in I - J \neq \varnothing$, we have by taking the inner product of (2) with d that

$$\sum_{i \in J} u_i \nabla g_i(\bar{x})^t d + \sum_{i \in I-J} u_i \nabla g_i(\bar{x})^t d = 0,$$

which yields a contradiction since the first term above is nonpositive and the second term above is strictly negative because $u_i > 0$ for at least one $i \in I - J \neq \varnothing$. Thus $u_0 > 0$, and so $\bar{x}$ is a KKT point. $\square$

5.20 Let $g(d) \equiv d^t d - 1 \leq 0$ be the nonlinear defining constraint. Then $\nabla g(\bar{d})^t d = 2\bar{d}^t d$. Hence G_1 is the set G' defined in the text, and so by Lemma 5.2.1, we have that $T \subseteq G_1$. Therefore, we need to show that $G_1 \subseteq T$. Let d be a nonzero vector from G_1. If $\bar{d}^t d < 0$, i.e., $\nabla g(\bar{d})^t d < 0$, then we readily have that $d \in D$ (see the proof of Lemma 4.2.4 for details), and hence, $d \in T$. Thus, suppose that $\bar{d}^t d = 0$. Then d is tangential to the sphere $d^t d \leq 1$. Hence, since $C_2 \bar{d} < 0$, and $C_1 d \leq 0$ with $d \neq \bar{d}$, there exists a sequence $\{d_k\}$ of feasible points $d_k \to \bar{d}$, $d_k \neq \bar{d}$, and $d_k^t d_k = 1$ such that $\dfrac{d}{\|d\|} = \lim_{k \to \infty} \dfrac{d_k - \bar{d}}{\|d_k - \bar{d}\|}$. Therefore, $d \in T$.

CHAPTER 6

LAGRANGIAN DUALITY AND SADDLE POINT OPTIMALITY
CONDITIONS

6.2 For the problem illustrated in Figure 4.13, a possible sketch of the perturbation function $v(y)$ and the set G are very much similar to that shown in Figure 6.1 (note that the upper envelope of G also increases with y, and only a partial view of G (from above) is shaded in Figure 6.1. Hence, as in Figure 6.1, there is no duality gap for this case.

6.3 Let the left-hand side of the inequality be given by $\phi(\hat{x}, \hat{y})$. Hence, we get

$$\sup_{y \in Y} \inf_{x \in X} \phi(x, y) = \phi(\hat{x}, \hat{y}) = \inf_{x \in X} \phi(x, \hat{y}) \leq \inf_{x \in X} \sup_{y \in Y} \phi(x, y). \quad \square$$

6.4 Let $y_\lambda = \lambda y_1 + (1 - \lambda)y_2$, where y_1 and $y_2 \in R^{m+\ell}$ and where $\lambda \in [0,1]$. We need to show that $v(y_\lambda) \leq \lambda v(y_1) + (1 - \lambda)v(y_2)$. For this purpose, let

$$X(y_1) = \{x : g_i(x) \leq y_{1i}, \ i = 1,...,m, \ h_i(x) = y_{1,m+i}, \ i = 1,...,\ell, \ x \in X\}$$
$$X(y_2) = \{x : g_i(x) \leq y_{2i}, \ i = 1,...,m, \ h_i(x) = y_{2,m+i}, \ i = 1,...,\ell, \ x \in X\}$$
$$X(y_\lambda) = \{x : g_i(x) \leq y_{\lambda i}, \ i = 1,...,m, \ h_i(x) = y_{\lambda,m+i}, \ i = 1,...,\ell, \ x \in X\}$$

$v(y_k) = f(x_k)$, where x_k optimizes (6.9) when $y = y_k$, for $k = 1, 2$, and let

$v(y_\lambda) = f(x^*)$, where x^* optimizes (6.9) when $y = y_\lambda$.

By the definition of the perturbation function $v(y)$, this means that

$x_k \in X(y_k)$ and $f(x_k) = \min\{f(x) : x \in X(y_k)\}$ for $k = 1, 2$, and

$x^* \in X(y_\lambda)$ and $f(x^*) = \min\{f(x) : x \in X(y_\lambda)\}$.

Under the given assumptions (the functions $g_i(x)$ are convex, the functions $h_i(x)$ are affine, and the set X is convex) we have from the definition of convexity that $x_\lambda = \lambda x_1 + (1 - \lambda)x_2 \in X(y_\lambda)$ for any $\lambda \in [0,1]$. But $f(x^*) = \min\{f(x) : x \in X(y_\lambda)\}$, and so $f(x^*) \leq f(x_\lambda)$, which together with the convexity of $f(x)$ implies that $v(y_\lambda) = f(x^*) \le f(x_\lambda) \le \lambda f(x_1) + (1 - \lambda)f(x_2) = \lambda v(y_1) + (1 - \lambda)v(y_2)$. This completes the proof. $\square$

6.5 The perturbation function from Equation (6.9) for Example 6.3.5 is given by

$$v(y) = \min\{-x_1 - x_2 : x_1 + 2x_2 \le 3 + y, \text{ with } x_1, x_2 \in \{0, 1, 2, 3\}\}.$$

Hence, by examining the different combinations of discrete solutions in the (x_1, x_2)-space, we get

$$v(y) = \begin{cases} \infty & \text{if } y < -3 \\ 0 & \text{if } -3 \le y < -2 \quad [\text{evaluated at } (x_1, x_2) = (0, 0)] \\ -1 & \text{if } -2 \le y < -1 \quad [\text{evaluated at } (x_1, x_2) = (1, 0)] \\ -2 & \text{if } -1 \le y < 0 \quad [\text{evaluated at } (x_1, x_2) = (2, 0)] \\ -3 & \text{if } 0 \le y < 2 \quad [\text{evaluated at } (x_1, x_2) = (3, 0)] \\ -4 & \text{if } 2 \le y < 4 \quad [\text{evaluated at } (x_1, x_2) = (3, 1)] \\ -5 & \text{if } 4 \le y < 6 \quad [\text{evaluated at } (x_1, x_2) = (3, 2)] \\ -6 & \text{if } y \ge 6 \quad [\text{evaluated at } (x_1, x_2) = (3, 3)] \end{cases}$$

Note that the optimal primal solution is given by $(\bar{x}_1, \bar{x}_2) = (3, 0)$ of objective value -3, which also happens to be the optimum to the underlying linear programming relaxation in which we restrict x_1 and x_2 to lie in $[0, 3]$, thus portending the existence of a saddle point solution. Indeed, for $\bar{u} = 1$, we see from Example 6.3.5 that $\theta(\bar{u}) = -3$, and so $(\bar{x}_1, \bar{x}_2, \bar{u})$ is a saddle point solution and there does not exist a duality gap in this example. Moreover, we see that

$$v(y) \ge 3 - y, \ \forall y$$

as in Equation (6.10) of Theorem 6.2.7, thus verifying the necessary and sufficient condition for the absence of a duality gap.

6.7 Denote $S \equiv conv\{x \in X : Dx = d\}$, and note that since X is a compact discrete set, we have that S is a polytope. Hence, for any linear function $f(x)$, we have $\min f(x) : Dx = d, x \in X\} = \min\{f(x) : x \in S\}$. Therefore, for each fixed $\pi \in R^m$, we get

$$\theta(\pi) = \min\{c^t x + \pi^t (Ax - b) : Dx = d, x \in X\}$$
$$= \min\{c^t x + \pi^t (Ax - b) : x \in S\}.$$

Now, consider the LP : $\min\{c^t x : Ax = b, \ x \in S\}$. Then, by strong duality for LPs, we get

$$\min\{c^t x : Ax = b, \ x \in S\} = \max_{\pi \in R^m} \ \min_{x \in S} \ \{c^t x + \pi^t (Ax - b)\} = \max_{\pi \in R^m} \theta(\pi). \quad (1)$$

This establishes the required result. Moreover, the optimal value of Problem DP is given by $v^* = \min\{c^t x : x \in S^*\}$, where

$$S^* \equiv conv\{x : Ax = b, Dx = d, x \in X\} \subseteq \{x : Ax = b, x \in S\}. \quad (2)$$

Thus, we get $v^* \geq \min\{c^t x : Ax = b, x \in S\}$, which yields from (1) that $v^* \geq \max_{\pi \in R^m} \theta(\pi)$, where a duality gap exists if this inequality is strict.

Therefore, the disparity in (2) potentially causes such a duality gap. $\square$

6.8 Interchanging the role of x and y as stated in the exercise for convenience, and noting Exercise 6.7 and Section 6.4, we have

$$v_1 = \max_{\mu} \overline{\theta}(\mu) = \max_{\mu} \min_{(x,y)} \{c^t x + \mu^t (x - y) : Ay = b, \ y \in Y, \ Dx = d, \ x \in X\}$$

and

$$v_2 = \max_{\pi} \theta(\pi) = \max_{\pi} \min_{x} \{c^t x + \pi^t (Ax - b) : Dx = d, \ x \in X\}.$$

Let $S_1 \equiv conv\{x \in X : Dx = d\}$ and $S_2 \equiv conv\{y \in Y : Ay = b\}$.
Then from Exercise 6.7 (see also Section 6.4), we have that

$$v_1 = \min\{c^t x : x \in S_1, \ y \in S_2, \ x = y\} \quad (1)$$

and

$$v_2 = \min\{c^t x : Ax = b, \ x \in S_1\}. \quad (2)$$

Hence, we get

$$v_2 = \min\{c^t x : Ay = b, \ x = y, \ x \in S_1\}$$

$\le \min\{c^t x : x \in S_1, y \in S_2, x = y\} = v_1$, where the inequality follows since $S_2 \subseteq \{y : Ay = b\}$. This proves the stated result, where (1) and (2) provide the required partial convex hull relationships.

6.9 First, we show that if $(d_u, d_v) = (0,0)$, then $(\overline{u}, \overline{v})$ solves Problem (D). Problem (D) seeks the maximum of a concave function $\theta(u,v)$ over $\{(u,v) : u \ge 0\}$, and so the KKT conditions are sufficient for optimality. To show that $(\overline{u}, \overline{v})$ is a KKT point for (D), we need to demonstrate that there exists a vector z_1 such that

$$-\nabla_u \theta(\overline{u}, \overline{v}) - z_1 = 0$$
$$-\nabla_v \theta(\overline{u}, \overline{v}) = 0$$
$$z_1^t \overline{u} = 0, \; z_1 \ge 0.$$

By assumption, we have $\nabla_v \theta(\overline{u}, \overline{v}) = h(\overline{x}) = 0$, and $\nabla_u \theta(\overline{u}, \overline{v}) = g(\overline{x})$. Moreover, since $d_u = 0$, we necessarily have $g(\overline{x}) \le 0$ and $g(\overline{x})^t \overline{u} = 0$. Thus, $z_1 = -g(\overline{x})$ solves the KKT system, which implies that $(\overline{u}, \overline{v})$ solves (D). (Alternatively, note from above that if $\overline{x}$ evaluates $\theta(\overline{u}, \overline{v})$, then the given condition implies that $\overline{x}$ is feasible to P with $\overline{u}^t g(\overline{x}) = 0$, and hence $(\overline{x}, \overline{u}, \overline{v})$ is a saddle point, and so by Theorem 6.2.5, $\overline{x}$ and $(\overline{u}, \overline{v})$ respectively solve P and D with no duality gap.)

Next, we need to show that if $(d_u, d_v) \ne (0,0)$, then (d_u, d_v) is a feasible ascent direction of $\theta(u,v)$ at $(\overline{u}, \overline{v})$. Notice that v is a vector of unrestricted variables, and by construction $d_{ui} \ge 0$ whenever $\overline{u}_i = 0$. Hence, (d_u, d_v) is a feasible direction at $(\overline{u}, \overline{v})$. To show that it is also an ascent direction, let us consider $\nabla \theta(\overline{u}, \overline{v})^t d$:

$$\nabla \theta(\overline{u}, \overline{v})^t d = \nabla_u \theta(\overline{u}, \overline{v})^t d_u + \nabla_v \theta(\overline{u}, \overline{v})^t d_v = g(\overline{x})^t \hat{g}(\overline{x}) + h(\overline{x})^t h(\overline{x})$$
$$= h(\overline{x})^t h(\overline{x}) + \sum_{i:\overline{u}_i > 0} g_i^2(\overline{x}) + \sum_{i:\overline{u}_i = 0} g_i(\overline{x}) \max\{0, \; g_i(\overline{x})\}.$$

All the foregoing terms are nonnegative and at least one of these is positive, for otherwise, we would have $(d_u, d_v) = (0,0)$.

54

Thus, $\nabla\theta(\bar{u},\bar{v})^t d > 0$. This demonstrates that (d_u, d_v) is an ascent direction of $\theta(u,v)$ at $(\bar{u},\bar{v})$. $\square$

In the given numerical example,

$$\theta(u_1, u_2) = \min\{x_1^2 + x_2^2 + u_1(-x_1 - x_2 + 4) + u_2(x_1 + 2x_2 - 8):$$
$$(x_1, x_2) \in R^2\}.$$

Iteration 1: $(u_1, u_2) = (0, 0)$.

At $(u_1, u_2) = (0, 0)$ we have $\theta(0, 0) = 0$, with $\bar{x}_1 = \bar{x}_2 = 0$. Thus, $d_1 = \max\{0, 4\} = 4$, and $d_2 = \max\{0, -8\} = 0$. Next, we need to maximize the function $\theta(u_1, u_2)$ from $(0, 0)$ along the direction $(4, 0)$. Notice that

$\theta[(0, 0) + \lambda(4, 0)] = \theta(4\lambda, 0)$
$= \min\{x_1^2 + x_2^2 + 4\lambda(-x_1 - x_2 + 4) : (x_1, x_2) \in R^2\}$
$= \min\{x_1^2 - 4\lambda x_1 : x_1 \in R\}$
$+\min\{x_2^2 - 4\lambda x_2 : x_2 \in R\} + 16\lambda$
$= -8\lambda^2 + 16\lambda,$

and $\max\{\theta(4\lambda, 0) : \lambda \geq 0\}$ is achieved at $\lambda^* = 1$. Hence, the new iterate is $(4, 0)$.

Iteration 2: $(u_1, u_2) = (4, 0)$.

At $(u_1, u_2) = (4, 0)$ we readily obtain that

$$\theta(4, 0) = \min\{x_1^2 + x_2^2 + 4(-x_1 - x_2 + 4) : (x_1, x_2) \in R^2\} = 8,$$

with $\bar{x}_1 = \bar{x}_2 = 2$. Thus, $d_1 = g_1(2, 2) = 0$, and $d_2 = \max\{0, -2\} = 0$. Based on the property of the dual problem, we conclude that at $(u_1, u_2) = (4, 0)$ the Lagrangian dual function $\theta(u_1, u_2)$ attains its maximum value. Thus $(\bar{u}_1, \bar{u}_2) = (4, 0)$ is an optimal solution to Problem D.

6.14 Let $\theta_1(v_0,v)$ be the Lagrangian dual function for the transformed problem. That is, $\theta_1(v_0,v) = \inf\{f(x) + v_0^t(g(x) + s) + v^t h(x) : (x,s) \in X'\}$.

The above formulation is separable in the variables x and s, which yields

$$\theta_1(v_0,v) = \inf\{f(x) + v_0^t g(x) + v^t h(x) : x \in X\} + \inf\{v_0^t s : s \ge 0\}.$$

Note that if $v_0 \ge 0$, then $\inf\{v_0^t s : s \ge 0\} = 0$, and otherwise, we get $\inf\{v_0^t s : s \ge 0\} = -\infty$. Therefore, the dual problem seeks the unconstrained maximum of $\theta_1(v_0,v)$, where

$$\theta_1(v_0,v) = \begin{cases} \inf\{f(x) + v_0^t g(x) + v^t h(x) : x \in X\} & \text{if } v_0 \ge 0 \\ -\infty & \text{otherwise.} \end{cases}$$

This representation of $\theta_1(v_0,v)$ shows that the two dual problems are equivalent (with $v_0 = u$).

6.15 For simplicity, we switch to the minimization of $f(x) = -3x_1 - 2x_2 - x_3$.

a. $\theta(u) = -4u_1 - 3u_2 + \min\{(-3 + u_1)x_1 +$
$$(-2 + 2u_1)x_2 + (-1 + u_2)x_3 : x \in X\}. \tag{1}$$

The set X has three extreme points $x_1 = (0, 0, 0)$, $x_2 = (1, 0, 0)$, and $x_3 = (0, 2, 0)$, and three extreme directions $d_1 = (0, 0, 1)$, $d_2 = (0, \frac{1}{2}, \frac{1}{2})$, and $d_3 = (\frac{1}{3}, 0, \frac{2}{3})$. Hence, for $\theta(u) > -\infty$, we must have (examining the extreme directions) that

$$u \in U \equiv \{(u_1,u_2) : u_2 \ge 1, \ 2u_1 + u_2 \ge 3, \ u_1 + 2u_2 \ge 5\}. \tag{2}$$

Hence, any $u \ge 0$ such that $u \in U$ will achieve the minimum in (1) at an extreme point, whence,

$$\theta(u) = \min\{-4u_1 - 3u_2, \ -3u_1 - 3u_2 - 3, \ -3u_2 - 4\}. \tag{3}$$

Putting (2) and (3) together and simplifying the conditions, we get

$$\theta(u) = \begin{cases} -4u_1 - 3u_2 & \text{if } u_1 \geq 3 \text{ and } u_2 \geq 1 \\ -3u_1 - 3u_2 - 3 & \text{if } \dfrac{1}{3} \leq u_1 \leq 3, \ u_2 \geq 1, \ 2u_1 + u_2 \geq 3, \text{ and } u_1 + 2u_2 \geq 5 \\ -3u_2 - 4 & \text{if } u_1 \leq \dfrac{1}{3} \text{ and } 2u_1 + u_2 \geq 3 \\ -\infty & \text{otherwise.} \end{cases}$$

b. In this case, we get

$$\theta(u) = -2u_1 - 3u_2 + \min_{x \in X}\{(-3 + 2u_1)x_1 + (-2 + u_1)x_2 + (-1 - u_1 + u_2)x_3\}$$

i.e.,

$$\theta(u) = -2u_1 - 3u_2 + \min\{(-3 + 2u_1)x_1 + (-2 + u_1)x_2 :$$
$$x_1 + 2x_2 \leq 4, (x_1, x_2) \geq 0\}$$
$$+ \min\{(-1 - u_1 + u_2)x_3 : x_3 \geq 0\}.$$

Noting that the extreme points of the polytope in the first minimization problem in (x_1, x_2) are $(0, 0)$, $(4, 0)$, and $(0, 2)$, and that the second minimization problem yields an optimal objective function value of zero if $-u_1 + u_2 \geq 1$ and goes to $-\infty$ otherwise, we get that

$$\theta(u) = -2u_1 - 3u_2 + \min\{0, \ -12 + 8u_1, \ -4 + 2u_1\} \text{ if } -u_1 + u_2 \geq 1,$$

and is $-\infty$ otherwise. Thus,

$$\theta(u) = \begin{cases} -2u_1 - 3u_2 & \text{if } u_1 \geq 2 \text{ and } -u_1 + u_2 \geq 1 \\ -12 + 6u_1 - 3u_2 & \text{if } u_1 \leq 4/3 \text{ and } -u_1 + u_2 \geq 1 \\ -4 - 3u_2 & \text{if } 4/3 \leq u_1 \leq 2 \text{ and } -u_1 + u_2 \geq 1 \\ -\infty & \text{otherwise.} \end{cases}$$

c. We can select those constraints to define X that will make the minimization over this set relatively easy, e.g., when the minimization problem decomposes into a finite number of simpler, lower dimensional, independent problems.

6.21 Let $\gamma = \inf\{f(x) : g(x) \le 0, \ h(x) = 0, \ x \in X\}$. Readily, γ is a finite number, since $\bar{x}$ solves Problem P: minimize $f(x)$ subject to $g(x) \le 0$, $h(x) = 0, \ x \in X$. Moreover, the system

$$f(x) - \gamma < 0, \ g(x) \le 0, \ h(x) = 0, \ x \in X$$

has no solution. By Lemma 6.2.3, it then follows that there exists a nonzero vector $(\bar{u}_0, \bar{u}, \bar{v})$, such that $(\bar{u}_0, \bar{u}) \ge 0$, and

$$\bar{u}_0(f(x) - \gamma) + \bar{u}^t g(x) + \bar{v}^t h(x) \ge 0 \text{ for all } x \in X. \tag{1}$$

That is, $\phi(\bar{u}_0, \bar{u}, \bar{v}, x) \ge \bar{u}_0 \gamma$ for all $x \in X$. But, since $\bar{x}$ solves Problem P, we have $\gamma = f(\bar{x})$. Moreover, $h(\bar{x}) = 0$ and $g(\bar{x}) \le 0$, so that $\bar{v}^t h(\bar{x}) = 0$ and $\bar{u}^t g(\bar{x}) \le 0$. Therefore, for any x in X

$$\phi(\bar{u}_0, \bar{u}, \bar{v}, x) \ge \bar{u}_0 f(\bar{x}) + \bar{u}^t g(\bar{x}) + \bar{v}^t h(\bar{x}) = \phi(\bar{u}_0, \bar{u}, \bar{v}, \bar{x}).$$

This establishes the second inequality. To prove the first inequality, note that for any $u \ge 0$, we have

$$\phi(\bar{u}_0, \bar{u}, \bar{v}, \bar{x}) - \phi(\bar{u}_0, u, v, \bar{x}) = (\bar{u} - u)^t g(\bar{x}) + $$
$$(\bar{v} - v)^t h(\bar{x}) = (\bar{u} - u)^t g(\bar{x}) \ge \bar{u}^t g(\bar{x}). \tag{2}$$

Now, from (1) for $x = \bar{x}$, since $f(\bar{x}) = \gamma$, we get $\bar{u}^t g(\bar{x}) + \bar{v}^t h(\bar{x}) \ge 0$, i.e., $\bar{u}^t g(\bar{x}) \ge 0$. But $g(\bar{x}) \le 0$ since $\bar{x}$ is a feasible solution, and $\bar{u} \ge 0$, which necessarily implies that $\bar{u}^t g(\bar{x}) = 0$. Thus, (2) implies that for any $u \ge 0$ and $v \in R^\ell$, we have that $\phi(\bar{u}_0, \bar{u}, \bar{v}, \bar{x}) - \phi(\bar{u}_0, u, v, \bar{x}) \ge 0$. $\square$

6.23 a. $\theta(u) = \min\{-2x_1 + 2x_2 + x_3 - 3x_4 + u_1(x_1 + x_2 + x_3 + x_4 - 8) + $
$$u_2(x_1 - 2x_3 + 4x_4 - 2) : x \in X\}$$
$$= \min\{-2 + u_1 + u_2)x_1 + (2 + u_1)x_2 : $$
$$x_1 + x_2 \le 8, x_1 \ge 0, x_2 \ge 0\}$$
$$+ \min\{1 + u_1 - 2u_2)x_3 + (-3 + u_1 + 4u_2)x_4 : $$
$$x_3 + 2x_4 \le 6, x_3 \ge 0, x_4 \ge 0\} - 8u_1 - 2u_2.$$

The extreme points of $\{(x_1,x_2) \geq 0 : x_1 + x_2 \leq 8\}$ are (0, 0), (8, 0), and (0, 8) in the (x_1,x_2)-space, and the extreme points of $\{(x_3,x_4) \geq 0 : x_3 + 2x_4 \leq 6\}$ are (0, 0), (6, 0), and (0, 3) in the (x_3,x_4)-space. Thus,

$$\theta(u) = -8u_1 - 2u_2 + \min\{0, -16 + 8u_1 + 8u_2, 16 + 8u_1\}$$
$$+ \min\{0, 6 + 6u_1 - 12u_2, -9 + 3u_1 + 12u_2\}. \qquad (1)$$

Noting that $(u_1,u_2) \geq 0$, we get that

$$\min\{0, -16 + 8u_1 + 8u_2, 16 + 8u_1\} = \begin{cases} 0 \text{ if } u_1 + u_2 \geq 2 \\ -16 + 8u_1 + 8u_2 \text{ if } u_1 + u_2 \leq 2. \end{cases} \qquad (2)$$

Similarly,

$$\min\{0, 6 + 6u_1 - 12u_2, 9 + 3u_1 + 12u_2\} =$$
$$\begin{cases} 0 \text{ if } -u_1 + 2u_2 \leq 1 \text{ and } u_1 + 4u_2 \geq 3 \\ 6 + 6u_1 - 12u_2 \text{ if } -u_1 + 2u_2 \geq 1 \text{ and} \\ \qquad\qquad -u_1 + 8u_2 \geq 5 \\ -9 + 3u_1 + 12u_2 \text{ if } u_1 + 4u_2 \leq 3 \text{ and} \\ \qquad\qquad -u_1 + 8u_2 \leq 5. \end{cases} \qquad (3)$$

Examining the six possible combinations given by (2) and (3), and incorporating these within (1), we get that (upon eliminating redundant conditions on (u_1,u_2)), $\theta(u) = \theta_i(u)$ if $u \in U_i$, $i = 1,\ldots,6$, where

$\theta_1(u) = -8u_1 - 2u_2$ and
$\quad U_1 = \{(u_1,u_2) \geq 0 : -u_1 + 2u_2 \leq 1, u_1 + u_2 \geq 2, u_1 + 4u_2 \geq 3\}$
$\theta_2(u) = 6 - 2u_1 - 14u_2$ and
$\quad U_2 = \{(u_1,u_2) \geq 0 : u_1 + u_2 \geq 2, -u_1 + 2u_2 \geq 1\}$
$\theta_3(u) = -9 - 5u_1 + 10u_2$ and
$\quad U_3 = \{(u_1,u_2) \geq 0 : u_1 + 4u_2 \leq 3, u_1 + u_2 \geq 2\}$
$\theta_4(u) = -16 + 6u_2$ and
$\quad U_4 = \{(u_1,u_2) \geq 0 : -u_1 + 2u_2 \leq 1, u_1 + u_2 \leq 2, u_1 + 4u_2 \geq 3\}$

$\theta_5(u) = -10 + 6u_1 - 6u_2$ and
$$U_5 = \{(u_1, u_2) \geq 0 : u_1 + u_2 \leq 2, -u_1 + 2u_2 \geq 1, -u_1 + 8u_2 \geq 5\}$$
$\theta_6(u) = -25 + 3u_1 + 18u_2$ and
$$U_6 = \{(u_1, u_2) \geq 0 : -u_1 + 8u_2 \leq 5, u_1 + 4u_2 \leq 3, u_1 + u_2 \leq 2\}$$

b. Note that $u = (4, 0)$ belongs to U_1 alone. Thus, θ is differentiable at $(4, 0)$, with $\nabla\theta(4,0) = (-8, -2)$.

c. When $u = (4, 0)$ and $d = (-8, -2)$, the second coordinate of $u + \lambda d = (4, 0) + \lambda(-8, -2)$ is -2λ, which is negative for all $\lambda > 0$. Since $u_2 = 0$, we have that the gradient of $\theta(u)$ at $(4, 0)$ is not a feasible direction at $(4, 0)$. However, projecting d onto $(d = d_2 \geq 0)$, we get that $d' = (-8, 0)$ is a feasible direction of $\theta(u)$ at $(4, 0)$. Moreoever, $\nabla\theta(4,0)^t d' = 64 > 0$. Thus, d' is an improving, feasible direction.

d. To maintain feasibility, we must have $4 - 8\lambda \geq 0$, i.e., λ should be restricted to values in the interval $[0, 1/2]$. Moreoever,

$$\theta(u + \lambda d') = \theta[(4,0) + \lambda(-8,0)] = \theta[(4 - 8\lambda, 0)]$$
$$= \min\{(2 - 8\lambda)x_1 + (6 - 8\lambda)x_2 : x_1 + x_2 \leq 8, x_1 \geq 0, x_2 \geq 0\}$$
$$+ \min\{(5 - 8\lambda)x_3 + (1 - 8\lambda)x_4 : x_3 + 2x_4 \leq 6, x_3 \geq 0, x_4 \geq 0\}$$
$$-32(1 - 2\lambda)$$
$$= -32(1 - 2\lambda) + \min\{0, 16(1 - 4\lambda), 16(3 - 4\lambda)\}$$
$$+ \min\{0, 6(5 - 8\lambda), 3(1 - 8\lambda)\}$$
$$= -32(1 - 2\lambda) + \min\{0, 16(1 - 4\lambda)\}$$
$$+ \min\{0, 3(1 - 8\lambda)\} \text{ when } 0 \leq \lambda \leq 1/2.$$

Thus, we get

$$\theta(u + \lambda d') = \begin{cases} -32 + 64\lambda & \text{for } 0 \leq \lambda < 1/8 \\ -29 + 40\lambda & \text{for } 1/8 \leq \lambda < 1/4 \\ -13 - 24\lambda & \text{for } 1/4 \leq \lambda \leq 1/2. \end{cases}$$

The maximum of $\theta(u + \lambda d') = \theta[(4 - 8\lambda, 0)]$ over $\lambda \in [0, 1/2]$ is -19, and is attained at $\lambda = 1/4$.

6.27 For any $u \geq 0$, we have

$$\theta(u) = \min_{x \geq 0}\{x + ug(x)\}.$$

a. For this case, we have (over $x \geq 0$):

$$x + ug(x) = \begin{cases} x - \dfrac{2u}{x} & \text{for } x > 0 \\ 0 & \text{for } x > 0. \end{cases}$$

When $u = 0$, we get $\theta(u) = 0$ (achieved uniquely at $x = 0$).

When $u > 0$, we get $\theta(u) = -\infty$ (as $x \to 0^+$).

Moreover, ξ is a subgradient of θ at $u = 0$ if and only if

$$\theta(u) \leq \theta(0) + u\xi, \ \forall u \geq 0$$

i.e., $\theta(u) \leq u\xi, \ \forall u \geq 0$.

Noting the form of θ, we get that any $\xi \in R$ is a subgradient. (Note that at $u = 0$, we get $\theta(u)$ is evaluated by only $x = 0$, where $g(0) = 0$, which is a subgradient, but Theorem 6.3.3 does not apply since g is not continuous at $x = 0$. Furthermore, if we consider all $u \in R$, then any $\theta(u) = 0$ for $u < 0$, and any $\xi \leq 0$ is a subgradient of θ at $u = 0$.)

b. For this case, we have (over $x \geq 0$) :

$$x + ug(x) = \begin{cases} x - \dfrac{2u}{x} & \text{for } x > 0 \\ -u & \text{for } x > 0. \end{cases}$$

When $u = 0$, we get $\theta(u) = 0$ evaluated (uniquely) at $x = 0$.

When $u > 0$, we get $\theta(u) = -\infty$ (as $x \to 0^+$).

As above, any $\xi \in R$ is a subgradient of θ at $u = 0$. Moreover, if we consider all of $u \in R$, then for $-8 \leq u \leq 0$, it can be verified that $\theta(u) = -u$ (see Part (c) below), so then any $\xi \leq -1$ is a subgradient of θ at $u = 0$.

c. In this case, we get

61

$$x + ug(x) = \begin{cases} x + \dfrac{2u}{x} & \text{for } x > 0 \\ u & \text{for } x = 0. \end{cases}$$

When $u = 0$, we get $\theta(u) = 0$, evaluated uniquely at $x = 0$.

When $u > 0$, we get $\theta(u) = \min\{u, \min_{x>0}\{x + \dfrac{2u}{x}\}\}$.

The convex function $x + \dfrac{2u}{x}$ over $x > 0$ achieves a minimum at $x = \sqrt{2u}$ of value $2\sqrt{2u}$. Hence, when $u > 0$, we get $\theta(u) = \min\{u, 2\sqrt{2u}\}$, i.e.,

$$\theta(u) = \begin{cases} 0 & \text{if } u = 0 \\ u & \text{if } u \le 8 \\ 2\sqrt{2u} & \text{if } u > 8. \end{cases}$$

Moreover, any $\xi \ge 1$ is a subgradient, considering either just $u \ge 0$ or all of $u \in R$, since in this case, $\theta(u) = -\infty$ when $u < 0$.

6.29 Assume that $X \ne \varnothing$.

a. The dual problem is: maximize $\theta(v)$, where $\theta(v) = \min\{f(x) + v^t(Ax - b) : x \in X\}$.

b. The proof of concavity of $\theta(v)$ is identical to that of Theorem 6.3.1. Alternatively, since X is a nonempty compact polyhedral set, and for each fixed v, since the function $f(x) + v^t(Ax - b)$ is concave, we have by Theorem 3.4.7 that there exists an extreme point of X that evaluates $\theta(v)$. Thus, if $vert(X)$ denotes the finite set of extreme points of X, we have that $\theta(v) = \min_{\hat{x} \in vert(X)} \{f(\hat{x}) + v^t(A\hat{x} - b)\}$. Thus, the dual function $\theta(v)$ is given by the minimum of a finite number of affine functions, and so is piecewise linear and concave [see also Exercise 3.9].

c. For a given $\hat{v}$, let $X(\hat{v})$ denote the set of optimal extreme point solutions to the problem of minimizing $f(x) + \hat{v}^t(Ax - b)$ over X.

Then, by Theorem 6.3.7, $\xi(\hat{v})$ is a subgradient of $\theta(v)$ at $\hat{v}$ if and only if $\xi(\hat{v}) = Ax - b$ for some x in the convex hull of $X(\hat{v})$. Moreoever, denoting $\partial\theta(\hat{v})$ as the subdifferential (set of subgradients) of θ at $\hat{v}$, we have that d is an ascent direction for θ at $\hat{v}$ if and only if $\inf\{\xi^t d : \xi \in \partial\theta(\hat{v})\} > 0$, i.e., if and only if $\min\{d^t(A\hat{x} - b) : \hat{x} \in X(\hat{v})\} > 0$. Hence, if $Ax = b$ for some $x \in X(\hat{v})$, then the set of ascent directions of $\theta(v)$ at v is empty. Otherwise, an ascent direction exists. In this case, the steepest ascent direction, $\hat{d}$, can be found by employing Theorem 6.3.11. Namely, $\hat{d} = \hat{\xi}/\|\hat{\xi}\|$, where $\hat{\xi}$ is a subgradient of $\theta(v)$ at $\hat{v}$ with the smallest Euclidean norm. To find $\hat{\xi}$, we can solve the following problem: minimize $\|Ax - b\|$ subject to $x \in conv[X(\hat{v})]$. If $\hat{x}$ is an optimal solution for this problem, then $\hat{\xi} = A\hat{x} - b$.

d. If X is not bounded, then it is not necessarily true that for each v there exists an optimal solution for the problem to minimize $f(x) + v^t(Ax - b)$ subject to $x \in X$. For all such vectors the dual function value $\theta(v)$ is $-\infty$. However, $\theta(v)$ is still concave and piecewise linear over the set of all vectors v for which $\min\{f(x) + v^t(Ax - b) : x \in X\}$ exists.

CHAPTER 7:

THE CONCEPT OF AN ALGORITHM

7.1 a. With $\alpha(x) = x^2$, and the map C defined in the exercise, we have:

if $x \in [-1, 1]$, then $y = C(x) = x$, and hence, $\alpha(y) = \alpha(x)$;

if $x < -1$, then $y = C(x) = x + 1$, and hence,

$$\alpha(y) = (x + 1)^2 < x^2 = \alpha(x);$$

if $x > 1$, then $y = C(x) = x - 1$, and hence,

$$\alpha(y) = (x - 1)^2 < x^2 = \alpha(x).$$

Evidently, the map B is closed over $(-\infty, 0) \cup (0, \infty)$, and moreover, if $x \neq 0$ and $y = B(x)$, then $y = \dfrac{x}{2}$, and hence $\alpha(y) = \dfrac{x^2}{4} < x^2 = \alpha(x)$. Thus, both the maps C and B satisfy all the assumptions of Theorem 7.3.4.

(b) For any x we have $A(x) = C(B(x)) = C(\dfrac{x}{2})$, which by the definition of the map C means that the map $A(x)$ is as given in Part (b) of this exercise. However, the composite map A is not closed at $x = -2$ and at $x = 2$. To see this, let $x_n = -2 - \dfrac{1}{n}$, $\forall n$. Then $x_n \to -2$, and $y_n = A(x_n) = -\dfrac{1}{2n}$ with $y_n \to 0$. But $0 \notin A(-2) = -1$. Similarly, let $x_n = 2 + \dfrac{1}{n}$, $\forall n$. Then $x_n \to 2$, and $y_n = A(x_n) = \dfrac{1}{2n}$ with $y_n \to 0$, but $0 \notin A(2) = 1$.

c. If the starting point $x_0 \in [-2, 2]$, then for each k we have $x_k \in [-2, 2]$, and thus $x_{k+1} = \dfrac{x_k}{2}$, which yields $\{x_k\} \to 0$. If the starting point $x_0 < -2$, then there exists an integer $K \geq 0$ such that if $k \leq K$, then $x_k < -2$, and if $k > K$, then $x_k \in [-2, 2]$. This means that except for a finite number of elements, all elements of the sequence $\{x_k\}$ are in the interval $[-2, 2]$, and thus from above, we have that for $k > K$, $x_{k+1} = \dfrac{x_k}{2}$, and so $\{x_k\} \to 0$. A similar argument can be used to show that if $x_0 > 2$, then we still obtain $\{x_k\} \to 0$.

7.2 a. A is closed for this case. To see this, let $\{x_n\} \to \bar{x}$ and let $\{y_n\} \to \bar{y}$ where $x_n^2 + y_n^2 \le 2$, $\forall n$. Thus in the limit, we get $\bar{x}^2 + \bar{y}^2 \le 2$. Hence, $\bar{y} \in A(\bar{x})$, and so A is closed.

b. Let $\{x_n\} \to \bar{x}$ and let $\{y_n\} \to \bar{y}$ where $x_n^t y_n \le 2$, $\forall n$. Thus, taking limits in the latter inequality, we get $\bar{x}^t \bar{y} \le 2$, and so $\bar{y} \in A(\bar{x})$. Thus, A is closed.

c. Let $\{x_n\} \to \bar{x}$ and let $\{y_n\} \to \bar{y}$ where $\left\| y_n - x_n \right\| \le 2$, $\forall n$. Hence, taking limits as $n \to \infty$, we get $\left\| \bar{y} - \bar{x} \right\| \le 2$, or that $\bar{y} \in A(\bar{x})$. Thus, A is closed.

d. Consider the sequences $\{x_n\} = \{\tfrac{1}{n}\}$ and $\{y_n\} = \left\{ \sqrt{1 - \dfrac{1}{n^2}} \right\}$. Then $\{x_n\} \to 0^+$, $\{y_n\} \to 1$, with $x_n^2 + y_n^2 = 1$, $\forall n$, i.e., $y_n \in A(x_n)$, $\forall n$. However, in the limit, $1 \notin A(0) \equiv [-1, 0]$. Thus A is not closed at $x = 0$.

7.3 Let Y denote the set $\{y : By = b,\ y \ge 0\}$. We need to show that for any sequence $\{x_k\}$ convergent to $\bar{x}$, if for each k, y_k is an optimal solution to the problem: minimize $x_k^t y$ subject to $y \in Y$, and $\{y_k\} \to \bar{y}$, then $\bar{y}$ solves the problem: minimize $\bar{x}^t y$ subject to $y \in Y$.

Note that for each k we have

$$y_k \in Y \tag{1}$$

and

$$x_k^t y_k \le x_k^t y \text{ for all } y \in Y. \tag{2}$$

Since Y is closed, (1) implies that $\bar{y} \in Y$. Moreover, taking the limit of the inequality in (2) as $k \to \infty$ for any fixed $y \in Y$, we get that $\bar{x}'\bar{y} \le \bar{x}'y$ for all $y \in Y$. Thus, $\bar{y}$ solves the problem to minimize $\{\bar{x}'y : y \in Y\}$. $\square$

7.6 It needs to be shown that for any sequences $\{x_k\}$ and $\{v_k\}$, if

1. $x_k \in X$, with $x_k \to \bar{x}$,

2. $v_k \in C(x_k), \forall k$, with $v_k \to \bar{v}$,

then $\bar{v} \in C(\bar{x})$. By the definition of the map C, $v \in C(x)$ for $x \in X$ means that there exist vectors $a \in A(x)$ and $b \in B(x)$ such that $v = a + b$. Hence, what needs to be shown is that under the given assumptions, we have $\bar{v} = \bar{a} + \bar{b}$, where $\bar{a} \in A(\bar{x})$, and $\bar{b} \in B(\bar{x})$, where $\bar{x} \in X$.

Since $v_k \in C(x_k), \forall k$, we have that for each k, there exist $a_k \in A(x_k)$ and $b_k \in B(x_k)$ such that $v_k = a_k + b_k$. Since Y is compact with $\{a_k\} \subseteq Y$ and $\{b_k\} \subseteq Y$, there exists a subsequence indexed by K such that $\{a_k\}_K \to \bar{a}$ and $\{b_k\}_K \to \bar{b}$. Since A and B are closed, we thus have that $\bar{a} \in A(\bar{x})$ and $\bar{b} \in B(\bar{x})$. Moreover, taking limits as $k \to \infty$, $k \in K$, we get $\{v_k\}_K \to \bar{a} + \bar{b}$. But the entire sequence $\{v_k\} \to \bar{v}$ by assumption. Hence, $\bar{v} = \bar{a} + \bar{b}$, where $\bar{a} \in A(\bar{x})$, $\bar{b} \in B(\bar{x})$, and $\bar{x} \in X$ (since X is closed).

7.7 We show that for any sequence $\{x_n, z_n\} \to (\bar{x}, \bar{z})$, and $y_n \in A(x_n, z_n)$ for each n with $\{y_n\} \to \bar{y}$, we have that $\bar{y} \in A(\bar{x}, \bar{z})$. The proof below is general for any vector norm in the sense that it does not use any specific type of vector norm.

By the definition of y_n, we have

$$y_n = \lambda_n x_n + (1 - \lambda_n)z_n, \text{ where } \lambda_n \in [0,1] \text{ is such that}$$
$$\|y_n\| \le \|\lambda x_n + (1 - \lambda)z_n\| \text{ for any } \lambda \in [0,1].$$

66

Note that the sequence $\{\lambda_n\}$ is bounded, and hence it must have a convergent subsequence. For notational simplicity and without loss of generality, assume that the sequence $\{\lambda_n\}$ itself converges, and let $\bar{\lambda}$ denote its limit. Then we can directly evaluate the limit $\bar{y}$ of $\{y_n\}$, that is $\bar{y} = \bar{\lambda}\bar{x} + (1 - \bar{\lambda})\bar{z}$. It remains to show that $\|\bar{y}\| \le \|\lambda\bar{x} + (1 - \lambda)\bar{z}\|$ for any $\lambda \in [0,1]$. For this purpose, note that for each n we have

$$\|y_n\| \le \|\lambda(x_n - \bar{x}) + (1 - \lambda)(z_n - \bar{z}) + \lambda\bar{x} + (1 - \lambda)\bar{z}\| \text{ for any } \lambda \in [0,1].$$

By the Cauchy-Schwartz property of vector norms $\|\cdot\|$, we thus obtain that, for each n, $\|y_n\| \le \|\lambda(x_n - \bar{x})\| + \|(1 - \lambda)(z_n - \bar{z})\| + \|\lambda\bar{x} + (1 - \lambda)\bar{z}\|$ for any $\lambda \in [0,1]$. Finally, by taking the limit as $n \to \infty$ we obtain

$$\|\bar{y}\| \le \|\lambda\bar{x} + (1 - \bar{\lambda})\bar{z}\| \text{ for any } \lambda \in [0,1], \text{ which completes the proof. } \square$$

7.8   We need to show that for any sequence $\{x_n, z_n\} \to (\bar{x}, \bar{z})$, if $\{y_n\} \to \bar{y}$, where for each n, y_n satisfies

$$\|y_n - x_n\| \le z_n \text{ and } \|y_n\| \le \|w\| \text{ for any } w \text{ such that } \|w - x_n\| \le z_n, \text{ then}$$

$$\|\bar{y} - \bar{x}\| \le \bar{z} \tag{1}$$
and
$$\|\bar{y}\| \le \|w\| \text{ for any } w \text{ such that } \|w - \bar{x}\| \le \bar{z}. \tag{2}$$

Since $\|y_n - x_n\| \le z_n$, $\forall n$, taking limits as $n \to \infty$, we have that (1) holds true. Hence, we need to establish (2). Note that we can assume that $z_n \ge 0$, $\forall n$. In case $\bar{z} = 0$, then (1) implies that $\bar{y} = \bar{x}$ and any w satisfying $\|w - \bar{x}\| \le \bar{z}$ must also yield $w = \bar{x}$, and so (2) holds. Thus, suppose that $\bar{z} > 0$. Now, by the definition of y_n, it follows that

$$y_n \text{ solves } \min\{\tfrac{1}{2}\|w\|^2 : \tfrac{1}{2}\|w - x_n\|^2 \le \tfrac{z_n^2}{2}\}, \ \forall n. \tag{3}$$

Since the KKT conditions are necessary and sufficient for the problem in (3) (with Slater's constraint qualification holding for n large

67

enough), we have that there exists a Lagrange multiplier u_n such that for each n, we get

$$y_n + u_n(y_n - x_n) = 0 \tag{4}$$

$$\|y_n - x_n\| \le z_n \tag{5}$$

$$u_n \ge 0, \ u_n\left[\|y_n - x_n\| - z_n\right] = 0. \tag{6}$$

Now, if $\bar{y} = \bar{x}$, then since $\bar{z} > 0$, (6) implies that in the limit, $\{u_n\} \to \bar{u} \equiv 0$. Otherwise, (4) yields that $\{u_n\} \to \bar{u}$ where $\bar{u} = \|\bar{y}\| / \|\bar{x} - \bar{y}\|$. In any case, the limit $\bar{u}$ of $\{u_n\}$ exists, and so taking limits in (4) – (6) as $n \to \infty$, we get

$$\bar{y} + \bar{u}(\bar{y} - \bar{x}) = 0, \ \|\bar{y} - \bar{x}\| \le \bar{z}, \ \bar{u} \ge 0, \ \bar{u}[\|\bar{y} - \bar{x}\| - \bar{z}] = 0. \tag{7}$$

This means that $\bar{y}$ satisfies the KKT conditions of

$$\min\{\tfrac{1}{2}\|w\|^2 : \tfrac{1}{2}\|w - \bar{x}\|^2 \le \frac{\bar{z}^2}{2}\}, \quad \text{i.e.,} \quad \bar{y} \quad \text{solves} \quad \text{this} \quad \text{(convex)}$$

program, and so (2) holds true. $\square$

7.19 See the proof of Lemma 5.1, and the comments that follow it in *Nonlinear Programming* by Willard I. Zangwill, Prentice Hall, Inc., Englewood Cliffs, NJ, 1969.

CHAPTER 8:

UNCONSTRAINED OPTIMIZATION

8.10 a. ($\Rightarrow$) Under the given assumptions, we have that:
there exists $\overline{\lambda} \in [a, b]$ such that $\theta(\overline{\lambda}) \leq \theta(\lambda)$ for all $\lambda \in [a, b]$,
and for any λ_1, $\lambda_2 \in [a, b]$ such that $\lambda_1 < \lambda_2$, we have:

(1) if $\lambda_2 \leq \overline{\lambda}$ then $\theta(\lambda_1) > \theta(\lambda_2)$

(2) if $\lambda_1 \geq \overline{\lambda}$ then $\theta(\lambda_1) < \theta(\lambda_2)$.

Let λ_1 and λ_2 be any two distinct values in the interval $[a, b]$.
Without loss of generality, assume that $\lambda_1 < \lambda_2$. Furthermore, let
$\hat{\lambda} \in (\lambda_1, \lambda_2)$. We need to show that $\theta(\hat{\lambda}) < \max\{\theta(\lambda_1), \theta(\lambda_2)\}$ for
any $\hat{\lambda} \in (\lambda_1, \lambda_2)$.

If $\lambda_2 \leq \overline{\lambda}$, then by (1) we have that $\max\{\theta(\lambda_1), \theta(\lambda_2)\} = \theta(\lambda_1)$.
Moreover, (1) yields $\theta(\hat{\lambda}) < \theta(\lambda_1)$ because $\lambda_1 < \hat{\lambda} < \overline{\lambda}$. Therefore,
$\theta(\hat{\lambda}) < \max\{\theta(\lambda_1), \theta(\lambda_2)\}$.

If $\lambda_1 \geq \overline{\lambda}$, then by (2) we have that $\max\{\theta(\lambda_1), \theta(\lambda_2)\} = \theta(\lambda_2)$, and
furthermore, (2) again yields $\theta(\hat{\lambda}) < \theta(\lambda_2)$ because $\lambda_2 > \hat{\lambda} > \overline{\lambda}$.

Finally, if $\lambda_1 < \overline{\lambda} < \lambda_2$, then $\hat{\lambda}$ must be either in the interval
$(\lambda_1, \overline{\lambda}]$ or else in the interval $(\overline{\lambda}, \lambda_2)$. In the latter case, we have
$\theta(\hat{\lambda}) < \theta(\lambda_2) \leq \max\{\theta(\lambda_1), \theta(\lambda_2)\}$ by the above property (2). In the
former case, the above property (1) yields $\theta(\hat{\lambda}) < \theta(\lambda_1) \leq$
$\max\{\theta(\lambda_1), \theta(\lambda_2)\}$. Therefore, $\theta(\hat{\lambda}) < \max\{\theta(\lambda_1), \theta(\lambda_2)\}$ for any
$\hat{\lambda} \in (\lambda_1, \lambda_2)$, which proves that $\theta(\lambda)$ is strongly quasiconvex over
the interval $[a, b]$.

($\Leftarrow$) Suppose that $\theta(\lambda)$ is strongly quasiconvex over the interval $[a,$
$b]$ and attains its minimum value over this interval at $\overline{\lambda}$. Note that the
minimizer $\overline{\lambda}$ is necessarily unique. Consider any two values λ_1 and
λ_2 in the interval $[a, b]$ such that $\lambda_1 < \lambda_2$. If $\lambda_2 = \overline{\lambda}$ or $\lambda_1 = \overline{\lambda}$,
then we readily have $\theta(\lambda_1) > \theta(\lambda_2)$ or $\theta(\lambda_2) > \theta(\lambda_1)$, respectively,

because $\bar{\lambda}$ is the unique minimizer of $\theta(\lambda)$ over the interval $[a, b]$. If $\lambda_1 > \bar{\lambda}$, then $\lambda_1 \in (\bar{\lambda}, \lambda_2)$, and $\theta(\lambda_1) < \max\{\theta(\bar{\lambda}), \theta(\lambda_2)\}$ $= \theta(\lambda_2)$. The last inequality follows from the assumed strong quasiconvexity of $\theta(\lambda)$, and the equality follows from the assumption that $\bar{\lambda}$ minimizes $\theta(\lambda)$ over $[a, b]$. If $\lambda_2 < \bar{\lambda}$, then $\lambda_2 \in (\lambda_1, \bar{\lambda})$, and by similar arguments we obtain $\theta(\lambda_2) < \max\{\theta(\lambda_1), \theta(\bar{\lambda})\}$ $= \theta(\lambda_1)$. This proves that $\theta(\lambda)$ is strongly unimodal over the interval $[a, b]$. $\square$

b.   The proof of this part follows basically the same type of arguments as in Part (a).

8.11 The quadratic function $q(\lambda) = a\lambda^2 + b\lambda + c$ passing through the points (λ_i, θ_i), $i = 1, 2, 3$, in the function-space implies that

$$a\lambda_1^2 + b\lambda_1 + c = \theta_1$$
$$a\lambda_2^2 + b\lambda_2 + c = \theta_2$$
$$a\lambda_3^2 + b\lambda_3 + c = \theta_3.$$

Solving for (a, b, c) yields the function as stated (which is in the form of a Lagrangian Interpolation Polynomial, and is readily verified to yield $q(\lambda_i) = \theta_i, \forall i = 1, 2, 3)$.

Furthermore, we have

$$q'(\lambda) = \frac{\theta_1(2\lambda - \lambda_2 - \lambda_3)}{(\lambda_1 - \lambda_2)(\lambda_1 - \lambda_3)} + \frac{\theta_2(2\lambda - \lambda_1 - \lambda_3)}{(\lambda_2 - \lambda_1)(\lambda_2 - \lambda_3)} + \frac{\theta_3(2\lambda - \lambda_1 - \lambda_2)}{(\lambda_3 - \lambda_1)(\lambda_3 - \lambda_2)} = 0 \quad (1)$$

when $\theta_1[2\lambda - (\lambda_2 + \lambda_3)](\lambda_2 - \lambda_3) + \theta_2[2\lambda - (\lambda_1 + \lambda_3)](\lambda_3 - \lambda_1) + \theta_3[2\lambda - (\lambda_1 + \lambda_2)](\lambda_1 - \lambda_2) = 0$.

Defining a_{ij} and b_{ij} as in the exercise, this yields

$$2\lambda[\theta_1 a_{23} + \theta_2 a_{31} + \theta_3 a_{12}] = \theta_1 b_{23} + \theta_2 b_{31} + \theta_3 b_{12}, \quad \text{i.e.,} \quad \lambda = \bar{\lambda} \quad \text{as}$$
defined in the exercise.

For the example, we have $\theta_1 = 4$, $\theta_2 = 1$, $\theta_3 = 7$, $a_{23} = -1$, $a_{31} = 3$, $a_{12} = -2$, $b_{23} = -7$, $b_{31} = 15$, and $b_{12} = -8$. Hence, we get

$$\bar{\lambda} = \frac{1}{2} \cdot \frac{(-7)(4) + (15)(1) + (-8)(7)}{(-1)(4) + (3)(1) + (-2)(7)} = 2.3.$$

In order to show that $\lambda_1 < \bar{\lambda} < \lambda_3$ when $(\lambda_1, \lambda_2, \lambda_3)$ satisfies the TPP, first of all, note that by differentiating (1), we get (upon performing some algebra) that

$$q''(\lambda) = \frac{2[(\theta_3 - \theta_2)(\lambda_2 - \lambda_1) + (\theta_1 - \theta_2)(\lambda_3 - \lambda_2)]}{(\lambda_2 - \lambda_1)(\lambda_3 - \lambda_1)(\lambda_3 - \lambda_2)}, \tag{2}$$

where the denominator in (2) is positive since $\lambda_1 < \lambda_2 < \lambda_3$, and the numerator is positive because of the TPP. Hence, $q(\lambda)$ is strictly convex, and so $\bar{\lambda}$ is its unique minimizer. Moreover, we get from (1) that (after some algebra)

$$q'(\lambda_1) = \frac{(\theta_2 - \theta_1)(\lambda_3 - \lambda_2)(\lambda_2 + \lambda_3 - 2\lambda_1) + (\theta_2 - \theta_3)(\lambda_2 - \lambda_1)^2}{(\lambda_2 - \lambda_1)(\lambda_3 - \lambda_1)(\lambda_3 - \lambda_2)}. \tag{3}$$

In (3), note that all the factors involving the λ-variables are positive and that $(\theta_2 - \theta_1) \leq 0$ and $(\theta_2 - \theta_3) \leq 0$, with at least one inequality strict by the TPP. Hence, $q'(\lambda_1) < 0$ and so q is strictly decreasing at λ_1. Likewise, it can be verified that $q'(\lambda_3) > 0$ and so q is strictly increasing at λ_3. Since $q'(\bar{\lambda}) = 0$ and q is strictly convex, we conclude that $\lambda_1 < \bar{\lambda} < \lambda_3$.

a. We can start with $\lambda_1 = 0$ and $\theta_1 = \theta(0)$. Assuming that the direction d defining $\theta(\lambda)$ is one of descent (else this procedure will determine a sequence for λ_2 that approaches zero), we can take a trial step $\hat{\lambda} > 0$ and compute $\hat{\theta} = \theta(\hat{\lambda})$.

Case (i): $\hat{\theta} \geq \theta_1$. In this case, we let $\lambda_3 = \hat{\lambda}$, $\theta_3 = \hat{\theta}$, and we continue bisecting the interval $[\lambda_1, \lambda_3]$, moving toward λ_1, until we find a bisection point $\tilde{\lambda}$ such that $\tilde{\theta} \equiv \theta(\tilde{\lambda}) < \theta_1$ (which exists since d is a descent direction). Then we set $\lambda_2 = \tilde{\lambda}$ and $\theta(\lambda_2) = \tilde{\theta}$, to obtain a TPP $(\lambda_1, \lambda_2, \lambda_3)$.

Case (ii): $\hat{\theta} < \theta_1$. In this case, we set $\lambda_2 = \hat{\lambda}$, $\theta_2 = \hat{\theta}$, and we continue doubling the interval $[\lambda_1, \lambda_2]$ until we find a right end-point $\tilde{\lambda}$ such that $\tilde{\theta} \equiv \theta(\tilde{\lambda}) \geq \theta_2$. If no such $\tilde{\lambda}$ exists, then θ is monotonically decreasing along d (not necessarily $\to -\infty$), and no TPP exists. Else, we set $\lambda_3 = \tilde{\lambda}$ and $\theta_3 = \tilde{\theta}$ to obtain a TPP $(\lambda_1, \lambda_2, \lambda_3)$.

b. Suppose that θ is strictly quasiconcave. By virtue of the TPP, there exists a minimizing solution $\lambda^* \in [\lambda_1, \lambda_3]$, and by Exercise 8.10b, θ is strictly unimodal over $[\lambda_1, \lambda_3]$.

Case (i): First consider the case when $\theta_1 \neq \theta_3$, where, without loss of generality, suppose that $\theta_1 < \theta_3$ (the case of $\theta_1 > \theta_3$ is similar). In this case, we get $\lambda_1 < \bar{\lambda} < \lambda_3$ from above, where by the strict quasiconvexity of θ, we have that $\bar{\theta} \equiv \theta(\bar{\lambda}) < \max\{\theta_1, \theta_3\} = \theta_3$. Now, consider the three cases discussed in Section 8.3.

Case 1: If $\bar{\lambda} > \lambda_2$, then the new TPP is $\lambda^1_{new} = (\lambda_1, \lambda_2, \bar{\lambda})$ if $\bar{\theta} \geq \theta_2$ and is $\lambda^2_{new} = (\lambda_2, \bar{\lambda}, \lambda_3)$ if $\bar{\theta} \leq \theta_2$. In the case of λ^1_{new}, if $\bar{\theta} > \theta_2$, then strict unimodality asserts that $\lambda^* \in [\lambda_1, \bar{\lambda}]$. Furthermore, if $\bar{\theta} = \theta_2$, then if $\theta_1 > \bar{\theta}$, we again get $\lambda^* \in [\lambda_1, \bar{\lambda}]$, and if $\theta_1 = \bar{\theta}$ also, then the function is constant over $[\lambda_1, \bar{\lambda}]$ with any point in this interval being an optimum. For the case of λ^2_{new}, since $\bar{\theta} < \theta_3$ and $\theta_2 \geq \bar{\theta}$, we again conclude the existence of $\lambda^* \in [\lambda_2, \lambda_3]$.

Case 2: If $\bar{\lambda} < \lambda_2$, the argument is similar to Case 1.

Case 3: If $\bar{\lambda} = \lambda_2$ and $\lambda_3 - \lambda_1 > \varepsilon$ (else we stop), we replace $\bar{\lambda}$ at a distance $\varepsilon/2$ away from λ_2 as described in Section 8.3, which yields Case 1 or Case 2 above.

Case (ii): $\theta_1 = \theta_3$. In this case, by the TPP, we must have $\theta_2 < \theta_1 = \theta_3$, and from above, we would obtain $\lambda_1 < \bar{\lambda} < \lambda_3$. By the strict quasiconvexity of θ, whether $\bar{\lambda} > \lambda_2$, $\bar{\lambda} < \lambda_2$, or $\bar{\lambda} = \lambda_2$, we must have $\bar{\theta} = \theta(\bar{\lambda}) < \theta_1 = \theta_3$. Now we can analyze the three cases of $\bar{\lambda} > \lambda_2$, $\bar{\lambda} < \lambda_2$, and $\bar{\lambda} = \lambda_2$ similar to above in order to conclude that the new TPP contains a minimizing solution.

c. Let $\theta(\lambda) = -3\lambda - 2\lambda^2 + 2\lambda^3 + 3\lambda^4$ for $\lambda \geq 0$. Hence, we get $\theta(0) = 0$ and $\theta(1) = 0$, and so we try $\lambda = 1/2$ to obtain $\theta(1/2) = -25/16 = -1.5625$. Thus, $(\lambda_1, \lambda_2, \lambda_3) = (0, 1/2, 1)$ yields a TPP, with $\theta_1 = 0$, $\theta_2 = -1.5625$, and $\theta_3 = 0$. For this case, we get $b_{31} = 1$ and $a_{31} = 1$. Thus, $\bar{\lambda} = \dfrac{1}{2} = \lambda_2$. Using $\varepsilon = 0.1$ for the sake of illustration, we replace $\bar{\lambda}$ at $0.5 + \varepsilon/2$, say, i.e., $\bar{\lambda} = 0.55$, which yields $\theta(0.55) = -1.6477 < \theta(\lambda_2)$. Thus, the new TPP is given by $(0.5, 0.55, 1)$ with the interval of uncertainty being 0.5. We can continue in this fashion until $\lambda_3 - \lambda_1 \leq \varepsilon$. The actual optimal solution for this instance is $\lambda^* = 0.638$ (to a tolerance of 0.001) with $\theta(\lambda^*) = -1.7116$.

8.12 a. We need to show that if $\theta_1 = \theta_2 = \theta_3 \equiv \theta$, then $\min\{\theta(\lambda) : \lambda \geq 0\} = \theta$.

By contradiction, suppose that $\min\{\theta(\lambda) : \lambda \geq 0\} = \theta(\lambda^*) < \theta$. Then by the strict quasiconvexity of $\theta(\lambda)$, since $\max\{\theta(\lambda^*), \theta(\lambda_i)\} = \theta(\lambda_i) = \theta$, $\forall i = 1, 2, 3$, we obtain:

(i) If $\lambda^* < \lambda_1$, then $\theta(\lambda) < \theta$, $\forall \lambda \in (\lambda^*, \lambda_2)$, which contradicts that $\theta(\lambda_1) = \theta$ because $\lambda_1 \in (\lambda^*, \lambda_2)$.

(ii) If $\lambda_1 < \lambda^* < \lambda_3$, then $\theta(\lambda) < \theta$, $\forall \lambda \in I \equiv (\lambda_1, \lambda^*) \cup (\lambda^*, \lambda_3)$, which contradicts that $\theta(\lambda_2) = \theta$ because $\lambda_2 \in I$.

(iii) If $\lambda^* > \lambda_3$, then $\theta(\lambda) < \theta$, $\forall \lambda \in (\lambda_2, \lambda^*)$, which contradicts that $\theta(\lambda_3) = \theta$ because $\lambda_3 \in (\lambda_2, \lambda^*)$.

Therefore, $\min\{\theta(\lambda) : \lambda \geq 0\} = \theta$. $\quad\square$

b. The continuity of $\bar{\theta}(\lambda)$, where $\lambda \equiv (\lambda_1, \lambda_2, \lambda_3)$, follows directly from the definition of continuity and the fact that θ is continuous. Hence, we need to show that $\bar{\theta}(\lambda_{new}) < \bar{\theta}(\lambda)$ whenever θ_1, θ_2, and θ_3 are not all equal to each other. The minimizer λ^* of the quadratic fit function lies in the interval (λ_1, λ_3). Let $\theta^* \equiv \theta(\lambda^*)$. First, suppose that $\lambda^* > \lambda_2$. If $\theta^* \geq \theta_2$, then $\lambda_{new} = (\lambda_1, \lambda_2, \lambda^*)$ and $\theta^* < \theta_3$ (by the strict quasiconvexity of θ and since $\theta_1, \theta_2, \theta_3$ are not all equal to each other), and so $\bar{\theta}(\lambda_{new}) = \bar{\theta}(\lambda) + \theta^* - \theta_3 < \bar{\theta}(\lambda) + \theta_3 - \theta_3 = \bar{\theta}(\lambda)$. If $\theta^* < \theta_2$, then $\lambda_{new} = (\lambda_2, \lambda^*, \lambda_3)$, and noting that $\theta^* < \theta_1$ since $\theta^* < \theta_2 \leq \theta_1$, we get $\bar{\theta}(\lambda_{new}) = \bar{\theta}(\lambda) + \theta^* - \theta_1 < \bar{\theta}(\lambda) + \theta_1 - \theta_1 = \bar{\theta}(\lambda)$. Hence, $\bar{\theta}(\lambda_{new}) < \bar{\theta}(\lambda)$ in this case. Similar derivations lead to $\bar{\theta}(\lambda_{new}) < \bar{\theta}(\lambda)$ if $\lambda^* < \lambda_2$, and likewise if $\lambda^* = \lambda_2$.

8.18 By the definition of $f(x) \equiv c^t x + \frac{1}{2} x^t Hx$ and $F(x)$, we have

$$F(x) = \nabla f(x)^t \nabla f(x) = (c + Hx)^t (c + Hx) = d^t x + \frac{1}{2} x^t Dx + c^t c,$$

where $d = 2Hc$ and $D = 2H^t H = 2H^2$.

If the steepest descent method is used to find the minimum of $F(x)$, then its rate of convergence is governed by the condition number of the matrix D. More precisely, the rate of convergence is bounded above by $(\beta - 1)^2/(\beta + 1)^2$, where β is the ratio of the largest to the smallest eigenvalue of D. Since $D = 2H^2$, the eigenvalues of D are two times the squares of the eigenvalues of H, and so $\beta = \alpha^2$, where α is the ratio of the largest to the smallest eigenvalue of matrix H. Next, simple algebra yields $(\alpha^2 - 1)^2/(\alpha^2 + 1)^2 > (\alpha - 1)^2/(\alpha + 1)^2$ whenever $\alpha > 1$. Since the smaller this ratio, the faster the convergence rate of the steepest descent method, this implies that the steepest descent method applied to the minimization of $F(x)$ will converge at a slower rate than when applied to minimizing $f(x)$.

8.19 Let $y \equiv K_k^2 \geq 0$, and noting that $\alpha > 1$, define $g(y) \equiv \dfrac{y}{(y + \alpha^3)(y + \alpha)}$.

We need to show that $g(y)$ is maximized at $y = \alpha^2$ over $y \geq 0$. Differentiating g, we get

$$g'(y) = \frac{\alpha^4 - y^2}{[(y + \alpha^3)(y + \alpha)]^2}.$$

Setting this to zero uniquely yields $y = \alpha^2$. Furthermore, we obtain

$$g''(y) = \frac{-2(y + \alpha^3)(y + \alpha)[\alpha^5 + \alpha^7 + y(3\alpha^4 - y^2)]}{[(y + \alpha^3)(y + \alpha)]^4}.$$

Hence, at least over $0 \leq y \leq \sqrt{3}\alpha^2$, g is concave, and so g is maximized at $y = \alpha^2$ over the interval $[0, \sqrt{3}\alpha^2]$. But for $y \geq \sqrt{3}\alpha^2$, we get $g'(y) < 0$, i.e., g is decreasing. Thus g is maximized at $y = \alpha^2$ over $y \geq 0$.

8.21 First, let us establish the Kantorovich inequality. Since the matrix H is symmetric and positive definite, there exist matrices Q and D, where Q is orthonormal and D is diagonal with positive diagonal entries given by the eigenvalues $0 < d_1 \leq d_2 \leq ... \leq d_n$ of H, such that $H = QDQ^t$. Now, conider the nonsingular linear transformation $x = Qy$. Then, noting that $Q^tQ = I$ since Q is orthonormal, i.e., $Q^{-1} = Q^t$, we get

$$x^t x = y^t Q^t Q y = y^t y$$
$$x^t H x = y^t Q^t [QDQ^t] Qy = y^t D y$$
and $\quad x^t H^{-1} x = y^t Q^t [QD^{-1}Q^t] Qy = y^t D^{-1} y.$

Thus the left-hand side of the Kantorovich inequality equals

$$f(y) \equiv \frac{(y^t y)^2}{(y^t D y)(y^t D^{-1} y)} \quad \text{for } y \in R^n.$$

Now, noting that $f(\alpha y) = f(y)$ for any $\alpha > 0$, we can restrict our attention without loss of generality to y such that $\|y\| = 1$, and so, we need to show equivalently that

$$\min_{\|y\|=1} \left[\frac{1}{(y^t Dy)(y^t D^{-1} y)} \right] \geq \frac{4\alpha}{(1+\alpha)^2}$$

i.e., $\max_{\|y\|=1} \left[(y^t Dy)(y^t D^{-1} y) \right] \leq \frac{(1+\alpha)^2}{4\alpha}.$ \hfill (1)

For any normalized vector y the expression $y^t Dy = \sum_{i=1}^{n} d_i y_i^2$ is simply a convex combination of $d_1, \ldots, d_n$, that is, it represents some value in the interval $[d_1, d_n]$, where d_1 and d_n are respectively the smallest and the largest eigenvalues of H. Therefore, for any fixed y, there exists a unique $\beta \in [0,1]$ such that $y^t Dy = \beta d_1 + (1 - \beta) d_n$, where by algebra, we have

that $\beta = \sum_i y_i^2 \frac{(d_n - d_i)}{(d_n - d_1)}$. Likewise, $y^t D^{-1} y = \hat{\beta}(\frac{1}{d_1}) + (1 - \hat{\beta})(\frac{1}{d_n})$

where $\hat{\beta} = \sum_i y_i^2 \dfrac{\left[\dfrac{1}{d_n} - \dfrac{1}{d_i} \right]}{\left[\dfrac{1}{d_n} - \dfrac{1}{d_1} \right]} = \sum_i y_i^2 \dfrac{(d_n - d_i)}{(d_n - d_1)} \cdot \dfrac{d_1}{d_i}.$

Hence, since $d_1/d_i \leq 1$, $\forall i$, we have that $\hat{\beta} \leq \beta$. Consequently, since $\frac{1}{d_1} \geq \frac{1}{d_n}$, we get that $y^t D^{-1} y \leq \beta(\frac{1}{d_1}) + (1 - \beta)(\frac{1}{d_n})$. Thus, for the left-hand side of (1), we get

$$\min_{\|y\|=1} \left[(y^t Dy)(y^t D^{-1} y) \right] \leq \max_{0 \leq \beta \leq 1} \left[\beta d_1 + (1 - \beta) d_2 \right] \left[\beta \left(\frac{1}{d_1} \right) + (1 - \beta) \left(\frac{1}{d_n} \right) \right]$$

$$= \max_{0 \leq \beta \leq 1} \left[-\frac{(\alpha - 1)^2}{\alpha} \beta^2 + \frac{(\alpha - 1)^2}{\alpha} \beta + 1 \right], \hfill (2)$$

76

where $\alpha \equiv d_n/d_1$. The concave function in the last maximand is maximized when $\beta = 1/2$ and yields the value $\dfrac{(\alpha - 1)^2}{4\alpha} + 1 = \dfrac{(\alpha + 1)^2}{4\alpha}$, and so from (2), we have that (1) holds true. This establishes the Kantorovich inequality.

For the bound on the convergence rate we then have for any $x \in R^n$,

$$1 - \frac{(x^t x)^2}{(x^t H x)(x^t H^{-1} x)} \le 1 - \frac{4\alpha}{(1 + \alpha)^2} = \frac{1 - 2\alpha + \alpha^2}{(1 + \alpha)^2} = \frac{(\alpha - 1)^2}{(\alpha + 1)^2},$$

which thus establishes (8.18). □

8.23 Author's note: This is a good problem to assign to the students as a project exercise since it illustrates well the numerical behavior of the different algorithmic schemes of Parts a – h, revealing the effect of ill-conditioning on these various methods. Observe that the optimal solution to this problem is given by (1, 1,..., 1), of objective value equal to zero.

8.27 Since the minimum of $f(x + \lambda d_j)$ over $\lambda \in R$ occurs at $\lambda = 0$, $\forall j = 1,...,n$, we have that

$$\frac{d}{d\lambda} f(x + \lambda d_j)\Big|_{\lambda=0} = \nabla f(x)^t d_j = 0, \ \forall j = 1,...,n,$$

i.e., $D\nabla f(x) = 0$, where D is an $n \times n$ matrix whose rows are given by $d_1^t,...,d_n^t$. Since $d_1,...,d_n$ are linearly independent, we have that D is nonsingular, which therefore implies that $\nabla f(x) = 0$. However, this does not imply that f has a local minimum at x (for example, see Figure 4.1 with $x = (0, 0)$, $d_1 = (1, 0)$, and $d_2 = (0, 1)$).

8.28 Since H is $n \times n$ and symmetric, it has n real eigenvalues $\lambda_1,...,\lambda_n$ with a corresponding set of n orthogonal (thus linearly independent) eigenvectors or characteristic vectors $d_1,...,d_n$, such that $Hd_i = \lambda_i d_i$, $\forall i = 1,...,n$. For any $i \ne j \in \{1,...,n\}$, we thus have $d_i^t H d_j = \lambda_j d_i^t d_j = 0$, and so $d_1,...,d_n$ are H-conjugate.

8.32 a. We establish this result by induction. First, consider d_1 and d_2, where $d_1 = a_1$ and $d_2 = a_2 - \left[\dfrac{d_1^t H a_2}{d_1^t H d_1} \right] d_1$. Then, we have

$$d_1^t H d_2 = d_1^t H a_2 - \left[\frac{d_1^t H a_2}{d_1^t H d_1} \right] \left(d_1^t H d_1 \right) = 0. \text{ Moreover, } d_1 \text{ and } d_2 \text{ are}$$

linearly independent since a_2 does not belong to the one-dimensional space spanned by $a_1 \equiv d_1$. Now, by induction, suppose that $d_1, \ldots, d_{k-1}$ are linearly independent with $d_i^t H d_j = 0$, $\forall i \neq j \in \{1, \ldots, k-1\}$ for some $3 \leq k \leq n$, and consider the case of d_k. Note that $d_1, \ldots, d_{k-1}$ lie in the $(k-1)$-dimensional space spanned by the linearly independent vectors $a_1, \ldots, a_{k-1}$, but since a_k does not belong to this space, we have that $d_1, \ldots, d_k$ are linearly independent. Therefore, to complete the proof, we need to show that $d_p^t H d_k = 0$, $\forall p = 1, \ldots, k-1$. Hence, consider any $p \in \{1, \ldots, k-1\}$. We then have

$$d_p^t H d_k = d_p^t H a_k - \sum_{i=1}^{k-1} \left(\frac{d_i^t H a_k}{d_i^t H d_i} \right) d_p^t H d_i$$

$$= d_p^t H a_k - \left(\frac{d_p^t H a_k}{d_p^t H d_p} \right) d_p^t H d_p$$

$$\left[\text{since } d_p^t H d_i = 0, \ \forall i \in \{1, \ldots, k-1\}, \ i \neq p \right].$$

This establishes the *H*-conjugancy result.

b. Since the vectors $a_1, \ldots, a_n$ are unit vectors in R^n, and each d_k is some linear combination of the vectors a_i for $1 \leq i \leq k$, w clearly have that D is upper triangular. Moreover, the only nonzero element in the *k*th row of d_k arises from a_k, which is therefore of unit value. Thus, D has unit elements along its diagonal.

c. For this instance, we get $d_1 = (1, 0, 0)^t$,

78

$$d_2 = \begin{bmatrix} 1 \\ -1 \\ 4 \end{bmatrix} - \begin{bmatrix} d_1^t H a_2 \\ d_1^t H d_1 \end{bmatrix} \begin{bmatrix} 1 \\ 0 \\ 0 \end{bmatrix} = \begin{bmatrix} 1 \\ -1 \\ 4 \end{bmatrix} - \left(\frac{-2}{2} \right) \begin{bmatrix} 1 \\ 0 \\ 0 \end{bmatrix} = \begin{bmatrix} 2 \\ -1 \\ 4 \end{bmatrix}.$$

$$d_3 = \begin{bmatrix} 2 \\ -1 \\ 6 \end{bmatrix} - \begin{bmatrix} d_1^t H a_3 \\ d_1^t H d_1 \end{bmatrix} \begin{bmatrix} 1 \\ 0 \\ 0 \end{bmatrix} - \begin{bmatrix} d_2^t H a_3 \\ d_2^t H d_2 \end{bmatrix} \begin{bmatrix} 2 \\ -1 \\ 4 \end{bmatrix},$$

i.e.,

$$d_3 = \begin{bmatrix} 2 \\ -1 \\ 6 \end{bmatrix} - \left(\frac{-2}{2} \right) \begin{bmatrix} 1 \\ 0 \\ 0 \end{bmatrix} - \left(\frac{19}{11} \right) \begin{bmatrix} 2 \\ -1 \\ 4 \end{bmatrix} = \begin{bmatrix} -5/11 \\ 8/11 \\ -10/11 \end{bmatrix}.$$

d. In this case, $a_1 = (1, 0, 0)^t$, $a_2 = (0, 1, 0)^t$, and $a_3 = (0, 0, 1)^t$.
Thus, we get $d_1 = (1, 0, 0)^t$,

$$d_2 = \begin{bmatrix} 0 \\ 1 \\ 0 \end{bmatrix} - \begin{bmatrix} d_1^t H a_2 \\ d_1^t H d_1 \end{bmatrix} \begin{bmatrix} 1 \\ 0 \\ 0 \end{bmatrix} = \begin{bmatrix} 0 \\ 1 \\ 0 \end{bmatrix} - \left(\frac{0}{2} \right) \begin{bmatrix} 1 \\ 0 \\ 0 \end{bmatrix} = \begin{bmatrix} 0 \\ 1 \\ 0 \end{bmatrix}.$$

$$d_3 = \begin{bmatrix} 0 \\ 0 \\ 1 \end{bmatrix} - \begin{bmatrix} d_1^t H a_3 \\ d_1^t H d_1 \end{bmatrix} \begin{bmatrix} 1 \\ 0 \\ 0 \end{bmatrix} - \begin{bmatrix} d_2^t H a_3 \\ d_2^t H d_2 \end{bmatrix} \begin{bmatrix} 0 \\ 1 \\ 0 \end{bmatrix}$$

$$= \begin{bmatrix} 0 \\ 0 \\ 1 \end{bmatrix} - \left(\frac{-1}{2} \right) \begin{bmatrix} 1 \\ 0 \\ 0 \end{bmatrix} - \left(\frac{2}{3} \right) \begin{bmatrix} 0 \\ 1 \\ 0 \end{bmatrix} = \begin{bmatrix} 1/2 \\ -2/3 \\ 1 \end{bmatrix}.$$

Note that $D = \begin{bmatrix} 1 & 0 & 1/2 \\ 0 & 1 & -2/3 \\ 0 & 0 & 1 \end{bmatrix}$ is upper triangular with ones along the
diagonal.

8.35 For notational simplicity, the subscript j is dropped, i.e., we let $D \equiv D_j$,
$\nabla f(y) \equiv \nabla f(y_j)$, $q \equiv q_j$, $p \equiv p_j$, $\lambda \equiv \lambda_j$, $\tau \equiv \tau_j$, and $v \equiv v_j$.
Furthermore, we let $\nabla f(y_+) \equiv \nabla f(y_{j+1})$ and $D_+ \equiv D_{j+1}$, and we denote
$a \equiv \nabla f(y)^t D \nabla f(y)$, $b \equiv \nabla f(y_+)^t D \nabla f(y_+)$, and $c \equiv q^t D q$. We need to
show that there exists a value for ϕ such that $D_+ \nabla f(y_+) = 0$. By
Equations (8.45) and (8.46), we have

$$D_+ \nabla f(y_+) = [D + \frac{pp^t}{p^t q} - \frac{Dqq^t D}{c} + \phi\tau \frac{vv^t}{p^t q}](q + \nabla f(y))$$

$$= Dq + \frac{pp^t q}{p^t q} - \frac{Dqq^t Dq}{c} + \phi\tau \frac{vv^t q}{p^t q} + D\nabla f(y) + \frac{pp^t \nabla f(y)}{p^t q} - \frac{Dqq^t D\nabla f(y)}{c}$$

$$+ \phi\tau \frac{vv^t \nabla f(y)}{p^t q}$$

$$= p + D\nabla f(y) + \frac{pp^t \nabla f(y)}{p^t q} - \frac{Dqq^t D\nabla f(y)}{c} + \phi\tau \frac{vv^t \nabla f(y)}{p^t q},$$

where the last step follows from $c = q^t Dq$, and $v^t q = 0$ (by virtue of the selection of τ). Furthermore, since $p^t q = -p^t \nabla f(y)$ (because $p^t \nabla f(y_+) = 0$ due to exact line searches), we obtain

$$D_+ \nabla f(y_+) = D\nabla f(y) - \frac{Dqq^t D\nabla f(y)}{c} + \phi\tau \frac{vv^t \nabla f(y)}{p^t q}. \tag{1}$$

By Equation (8.47) and the definition of $v \equiv p - \frac{Dq}{\tau}$, we get

$$\frac{\tau vv^t \nabla f(y)}{p^t q} = \frac{cpp^t \nabla f(y)}{(p^t q)^2} - \frac{pq^t D\nabla f(y)}{p^t q} - \frac{Dqp^t \nabla f(y)}{p^t q} + \frac{Dqq^t D\nabla f(y)}{c}. \tag{2}$$

Now, as per the Hint, and due to exact line searches, let us note the following relationships:

$$pp^t \nabla f(y) = \lambda^2 aD\nabla f(y), \quad p^t q = \lambda a, \quad q^t D\nabla f(y) = -q^t p/\lambda = -a,$$
$$pq^t D\nabla f(y) = -ap, \text{ and}$$
$$Dqp^t \nabla f(y) = -\lambda Dq\nabla f(y)^t D\nabla f(y) = -\lambda aDq. \tag{3}$$

Based on (3), we get from Equation (2) that

$$\tau \frac{vv^t \nabla f(y)}{p^t q} = \frac{c}{a} D\nabla f(y) - D\nabla f(y) + Dq - \frac{a}{c} Dq.$$

Note that $c = a + b$ since $\nabla f(y)^t D\nabla f(y_+) = 0$, which results in

$$\tau \frac{vv^t \nabla f(y)}{p^t q} = \frac{b}{a} D\nabla f(y) + \frac{b}{c} Dq.$$ Substituting this in (1), we get using (3),

$$D_+ \nabla f(y_+) = D\nabla f(y) + \frac{a}{c} Dq + \phi[\frac{b}{a} D\nabla f(y) + \frac{b}{c} Dq]$$

$$= (1 + \phi\frac{b}{a})D\nabla f(y) + \frac{a}{c}(1 + \phi\frac{b}{a})Dq = (1 + \phi\frac{b}{a})D(\nabla f(y) + \frac{a}{c}q).$$

This demonstrates that if $\phi = -\frac{a}{b} = -\dfrac{\nabla f(y)^t D\nabla f(y)}{\nabla f(y_+)^t D\nabla f(y_+)}$, then $D_+\nabla f(y_+)$

is automatically zero, thus completing the proof. $\square$

8.41 We derive C by solving the following problem:

> **P:** Minimize $\dfrac{1}{2}\sum_i \sum_j C_{ij}^2$
>
> subject to
> $$[C + B_k]p_k = q_k$$
> $$C_{ij} = C_{ji} \quad \forall i, j.$$

Let $b \equiv q_k - B_k p_k$, and let $p_k \equiv [a_1,...,a_n]^t$. Substituting this into Problem P, and factoring in the symmetry condition by defining C_{ij} only for $i \le j$, we get:

> **P:** Minimize $\dfrac{1}{2}\sum_{i=1}^n C_{ii}^2 + \sum_{i=1}^{n-1}\sum_{j=i+1}^n C_{ij}^2$
>
> subject to
> $$C_{ii}a_i + \sum_{j<i} C_{ji}a_j + \sum_{j>i} C_{ji}a_j = b_i, \quad \forall i = 1,...,n.$$

Because P is a linearly constrained convex program, the KKT conditions are both necessary and sufficient for optimality. Denoting $\lambda_1,...,\lambda_n$ as the (unrestricted) Lagrange multipliers associated with the respective equality constraints, the KKT conditions yield

$$C_{ii} - \lambda_i a_i = 0, \quad \forall i = 1,...,n \tag{1}$$

$$2C_{ij} - a_j \lambda_i - a_i \lambda_j = 0, \ \forall i < j \tag{2}$$

$$C_{ii} a_i + \sum_{j<i} C_{ij} a_j + \sum_{j>i} C_{ij} a_j = b_i, \ \forall i = 1,...,n. \tag{3}$$

From (1) and (2), we get

$$C_{ii} = a_i \lambda_i, \ \forall i = 1,...,n \tag{4}$$

$$C_{ij} = \frac{a_i \lambda_j + a_j \lambda_i}{2}, \ \forall i < j. \tag{5}$$

Substituting (4) and (5) into (3) and collecting terms yields

$$\lambda_i \left[a_i^2 + \frac{1}{2} \sum_{j \neq i} a_j^2 \right] + \frac{a_i}{2} \sum_{j \neq i} a_j \lambda_j = b_i, \ \forall i = 1,...,n. \tag{6}$$

Define

$$Q = \frac{1}{2} \| p_k \|^2 I + \frac{1}{2} p_k p_k^t. \tag{7}$$

Note that Q is symmetric and positive definite (since $p_k \neq 0$), and is therefore nonsingular. Moreover, (6) is given by $Q\lambda = b$. Thus, we get

$$\lambda = \left[\frac{1}{2} \| p_k \|^2 I + \frac{1}{2} p_k p_k^t \right]^{-1} \left[q_k - B_k p_k \right],$$

or using the Sherman-Morrison-Woodbury formula (see Equation (8.55)), we get

$$\lambda = \frac{2}{\| p_k \|^2} \left[I - \frac{1}{2} \frac{p_k p_k^t}{\| p_k \|^2} \right] \left[q_k - B_k p_k \right]. \tag{8}$$

Hence, the correction matrix C is given by (4) and (5), with $C_{ij} = C_{ji}, \forall i < j$, where λ is computed via (8).

8.47 Consider the following problem:

P: Minimize $x_1^2 + x_2^2$

         subject to $x_1 + x_2 - 4 = 0$.

a.   Since Problem P is a linearly constrained convex program, note that the KKT conditions are both necessary and sufficient for optimality. The KKT conditions for Problem P yield (with v unrestricted):

$$2x_1 - v = 0$$
$$2x_2 - v = 0$$
$$x_1 + x_2 = 4,$$

or that $x_1 = x_2 = 2$ and $v = 4$. Hence $x = (2, 2)^t$ solves Problem P.

b.   Consider the following (penalty) problem:

PP: Minimize $\{F(x) \equiv x_1^2 + x_2^2 + \mu(x_1 + x_2 - 4)^2\}$,

where $\mu > 0$. For $\mu = 10$, we get

$$\nabla F(x) = \begin{bmatrix} 2x_1 + 20(x_1 + x_2 - 4) \\ 2x_2 + 20(x_1 + x_2 - 4) \end{bmatrix}$$

Starting with $x^1 = (0, 0)^t$ and using Fletcher and Reeves' conjugate gradient method, we get:

(i) $y^1 = (0, 0)^t$, $d^1 = -\nabla F(y^1) = \begin{bmatrix} 80 \\ 80 \end{bmatrix}$, and λ_1 is given by

$\underset{\lambda \geq 0}{\text{minimize}} \ F(y^1 + \lambda d^1) = \underset{\lambda \geq 0}{\text{minimize}} \ \left[2(80\lambda)^2 + 10[160\lambda - 4]^2 \right]$.

Hence $\lambda_1 = 0.0249843$, and $y^2 = y^1 + \lambda_1 d^1 = \begin{bmatrix} 1.998744 \\ 1.998744 \end{bmatrix}$.

(ii)   Next, we compute $d^2 = -\nabla F(y^2) + \alpha_1 d^1$, where

$\alpha_1 = \dfrac{\left\| \nabla F(y^2) \right\|^2}{\left\| \nabla F(y^1) \right\|^2}$. Note that $\nabla F(y^2) = \begin{bmatrix} 3.947248 \\ 3.947248 \end{bmatrix}$, which

yields $\alpha_1 = \dfrac{31.161532}{12800}$. Thus,

$d^2 = \begin{bmatrix} -3.947248 \\ -3.947248 \end{bmatrix} + \dfrac{31.161532}{12800} \begin{bmatrix} 80 \\ 80 \end{bmatrix}$, i.e.,

$d^2 = \begin{bmatrix} -3.7524885 \\ -3.7524885 \end{bmatrix}$. We hence compute λ_2 by minimizing

$F(y^2 + \lambda d^2)$ over $\lambda \geq 0$. Solving $\dfrac{d}{d\lambda} F(y^2 + \lambda d^2) = 0$ yields

$\lambda_2 = 0.0250452$, which gives $y^3 = y^2 + \lambda_2 d^2 = \begin{bmatrix} 1.9047622 \\ 1.9047622 \end{bmatrix}$.

This completes one loop of the algorithm, and we get $x^2 = \begin{bmatrix} 1.9047622 \\ 1.9047622 \end{bmatrix}$, reset $y^1 \equiv x^2$ and repeat. Continuing in this fashion, we find an ε-optimal solution to Problem PP within some tolerance $\varepsilon > 0$ based on terminating the procedure when $\left\| \nabla F(y^j) \right\| < \varepsilon$ for some iterate j, where the true optimum is given by

$\begin{bmatrix} 1.9047619 \\ 1.9047619 \end{bmatrix}$.

8.51 a. Note that the points $\{x_1, x_1 + d_j \text{ for } j = 1,...,n\}$ define vertices of a simplex if they are affinely independent, i.e., if $d_1,...,d_n$ are linearly independent. To show this, consider the system of linear equations defined by $\sum\limits_{j=1}^{n} d_j \lambda_j = 0$. This yields

$$a\lambda_i + \sum\limits_{j \neq i} b\lambda_j = 0, \ \forall i = 1,...,n. \tag{1}$$

Performing row operations on (1) by subtracting the $(i+1)^{th}$ equation from the ith equation for $i = 1,...,n-1$ yields the following equivalent system of equations:

$$(a - b)\lambda_i + (b - a)\lambda_{i+1} = 0, \ \forall i = 1,...,n-1 \tag{2}$$

$$b\sum\limits_{j=1}^{n-1} \lambda_j + a\lambda_n = 0. \tag{3}$$

Since $(a - b) = c/\sqrt{2} > 0$, Equation (2) yields $\lambda_1 = \lambda_2 = \cdots = \lambda_n$, which when substituted into (3) implies that $\lambda_1 = \lambda_2 = \cdots = \lambda_n = 0$. Hence, $d_1,...,d_n$ are linearly independent.

Furthermore, observe that

$$\left\| d_j \right\|^2 = a^2 + (n-1)b^2 = \frac{c^2}{2n^2} \left[\sqrt{n+1} + n - 1 \right]^2 +$$

84

$$\frac{(n-1)c^2}{2n^2}\left[\sqrt{n+1}-1\right]^2 = c^2.$$

Thus, c represents the distance from x_1 to each of the other vertices $x_2,...,x_{n+1}$. Furthermore,

$$\left\|x_{r+1} - x_{s+1}\right\|^2 = \left\|d_r - d_s\right\|^2 = 2(a-b)^2 = c^2, \qquad \forall r \neq s \in \{1,...,n\}.$$

Hence c represents the length of each edge of the constructed simplex.

b. Consider the problem to minimize
$$f(x) = 2x_1^2 + 2x_1x_2 + x_3^2 + 3x_2^2 - 3x_1 - 10x_3.$$

Denoting the iterates with superscripts, and noting that $n = 3$, and taking $c = 3\sqrt{2}$, we get $a = 4$ and $b = 1$. Taking $x^1 = (0, 0, 0)^t$, this yields $x^2 = (4, 1, 1)^t$, $x^3 = (1, 4, 1)^t$, and $x^4 = (1, 1, 4)^t$. (Note that the length of each edge of this simplex is $c = 3\sqrt{2}$.) We then get:

Step 1: $f(x^1) = 0$, $f(x^2) = 22$, $f(x^3) = 46$, and $f(x^4) = -20$. Thus, $x^r \equiv x^4$ and $x^s \equiv x^3$, which yields

$$\bar{x} = \frac{1}{3}\left[x^1 + x^2 + x^4\right] = \left[5/3, \ 2/3, \ 5/3\right]^t.$$

Step 2: Using $\alpha = 1$, we get

$$\hat{x} = \begin{bmatrix} 5/3 \\ 2/3 \\ 5/3 \end{bmatrix} + \begin{bmatrix} 2/3 \\ -10/3 \\ 2/3 \end{bmatrix} = \begin{bmatrix} 7/3 \\ -8/3 \\ 7/3 \end{bmatrix}, \text{ with } f(\hat{x}) = \frac{-46}{9} > f(x^r),$$

and so we go to Step 4.

Step 4: Since $\max\{f(x^1), f(x^2), f(x^4)\} = 22 > f(\hat{x})$, we replace x^3 with $\hat{x}$ to get a simplex defined by $\{x^1, x^2, x^4 \ \hat{x}\}$ with $\hat{x}$ now playing the role of the revised x^3 (i.e., $x^3 \leftarrow \hat{x}$), and we return to Step 1 for the next iteration.

The iterations now continue until the simplex shrinks close to a point.

8.52 For notational simplicity let $D \equiv D_j$, and $d \equiv D\nabla f(y_j)$. Then $y_{j+1} = y_j - \lambda_j d$, where λ_j minimizes $f(y_j - \lambda d)$ along the direction $-d$. Note that

$$f(y_j - \lambda d) = c^t(y_j - \lambda d) + \frac{1}{2}(y_j - \lambda d)^t H(y_j - \lambda d)$$

$$= \frac{1}{2}\lambda^2 d^t H d - \lambda(c + Hy_j)^t d + c^t y_j + \frac{1}{2}y_j^t H y_j$$

$$= \frac{1}{2}\lambda^2 d^t H d - \lambda d^t D^{-1} d + c^t y_j + \frac{1}{2}y_j^t H y_j,$$

since $(c + Hy_j) = \nabla f(y_j) = D^{-1}d$, where D is assumed to be symmetric and positive definite. Thus, the minimizing value of λ is given by

$$\lambda_j = \frac{d^t D^{-1} d}{d^t H d}. \tag{1}$$

Next, we examine $e(y_{j+1})$, where y^* is a minimizing solution for f:

$$2e(y_{j+1}) = (y_{j+1} - y^*)^t H(y_{j+1} - y^*) = (y_{j+1} - y_j)^t H(y_{j+1} - y_j)$$
$$+2(y_{j+1} - y_j)^t H(y_j - y^*) + (y_j - y^*)^t H(y_j - y^*)$$
$$= \lambda_j^2 d^t H d - 2\lambda_j d^t D^{-1} d + 2e(y_j),$$

since $(y_{j+1} - y_j) = -\lambda_j d$ and $H(y_j - y^*) = Hy_j + c = D^{-1}d$ as above. Hence,

$$\frac{e(y_{j+1})}{e(y_j)} = 1 - \frac{\left[\lambda_j d^t D^{-1} d - \lambda_j^2 \left(\frac{1}{2} d^t H d\right)\right]}{e(y_j)}. \tag{2}$$

Note that

$$2e(y_j) = (y_j - y^*)^t H(y_j - y^*) = d^t D^{-1} H^{-1} D^{-1} d \tag{3}$$

because $\nabla f(y^*) = Hy^* + c = 0$, and so, $H(y_{j+1} - y^*) = c + Hy_j = \nabla f(y_j) = D^{-1}d$, which yields $(y_j - y^*) = H^{-1}D^{-1}d$ or $(y_j - y^*) = d^t D^{-1}H^{-1}$. Substituting Equations (1) and (3) into (2) yields

$$e(y_{j+1}) = e(y_j)\left[1 - \frac{(d^t D^{-1}d)^2}{(d^t Hd)(d^t D^{-1}H^{-1}D^{-1}d)}\right]. \tag{4}$$

By assumption, the matrix D is symmetric and positive definite. Therefore, there exist an orthonormal matrix Q and a diagonal positive definite matrix P (with the square-root of the eigenvalues of D along the diagonal) such that $D = QPPQ^t$. In order to express Equation (4) in a more convenient form, let $r \equiv P^{-1}Q^t d$ and let $G \equiv PQ^t HQP$. Then, since $QQ^t = I$, we get

$$d^t D^{-1}d = d^t QP^{-1}P^{-1}Q^t d = r^t r,$$
$$d^t HD = d^t QP^{-1}PQ^t HQPP^{-1}Q^t d = r^t PQ^t HQ\Pr = r^t Gr,$$

and $\quad d^t D^{-1}H^{-1}D^{-1}d = r^t P^{-1}Q^t H^{-1}QP^{-1}r = r^t G^{-1}r,$

which allows us to rewrite Equation (4) as

$$e(y_{j+1}) = e(y_j)\left[1 - \frac{(r^t r)^2}{(r^t Gr)(r^t G^{-1}r)}\right].$$

The matrix G is readily verified to be positive definite, and so the Kantorovich inequality (see Exercise 8.21 and Equation (8.18)) can be applied to obtain an upper bound on the convergence rate as

$$e(y_{j+1}) \le \frac{(\beta - 1)^2}{(\beta + 1)^2} e(y_j),$$

where β is the ratio of the largest to the smallest eigenvalue of the matrix G.

It remains to show that the matrices G and DH have the same eigenvalues. For this purpose note that if λ is an eigenvalue of DH and x is a corresponding eigenvector, then $DHx = \lambda x$, i.e., $QPPQ^t Hx = \lambda x$, which yields $PQ^t HQPz = \lambda z$, where $z \equiv P^{-1}Q^t x$. This demonstrates that

$Gz = \lambda z$, i.e., λ is an eigenvalue of G. The converse is likewise true. Therefore, $\beta = \alpha_j$, which completes the proof.

CHAPTER 9:

PENALTY AND BARRIER FUNCTIONS

9.2 Using the quadratic penalty function, the penalty problem is given as follows:

PP: $\underset{(x_1,x_2)}{\text{Minimize}}\{F_\mu(x) \equiv 2e^{x_1} + 3x_1^2 + 2x_1x_2 + 4x_2^2 + \mu[3x_1 + 2x_2 - 6]^2\}.$

Note that

$$\nabla F_\mu(x) = \begin{bmatrix} 2e^{x_1} + 6x_1 + 2x_2 + 6\mu[3x_1 + 2x_2 - 6] \\ 2x_1 + 8x_2 + 4\mu[3x_1 + 2x_2 - 6] \end{bmatrix}.$$

Hence, with $\mu = 10$, we get

$$F_\mu(x) \equiv 2e^{x_1} + 3x_1^2 + 2x_1x_2 + 4x_2^2 + 10[3x_1 + 2x_2 - 6]^2$$

and

$$\nabla F_\mu(x) = \begin{bmatrix} 2e^{x_1} + 6x_1(1 + 3\mu) + 2x_2(1 + 6\mu) - 36\mu \\ 2x_1(1 + 6\mu) + 8x_2(1 + \mu) - 24\mu \end{bmatrix}$$

$$= \begin{bmatrix} 2e^{x_1} + 186x_1 + 122x_2 - 360 \\ 122x_1 + 88x_2 - 240 \end{bmatrix}.$$

Using the conjugate gradient method of Fletcher and Reeves with $x = (1, 1)^t$ as the starting solution, we perform the following two iterations:

Iteration 1:

$$x^1 = (1, 1)^t, \quad \nabla F_\mu(x^1) = \begin{bmatrix} 2e - 52 \\ -30 \end{bmatrix}, \quad d^1 = -\nabla F_\mu(x^1) = \begin{bmatrix} 52 - 2e \\ 30 \end{bmatrix}$$

$= \begin{bmatrix} 46.563436 \\ 30 \end{bmatrix}$, and the step length λ is obtained as a solution to $\underset{\lambda \geq 0}{\text{minimize}}\{F_\mu(x^1 + \lambda d^1)\}$. The optimal step length is given by $\lambda^* = 0.004$, which yields the new iterate

$x^2 = x^1 + \lambda^* d^1 = [1.86253744, \ 1.12]^t.$

Iteration 2:

$$\nabla F_\mu(x^2) = \begin{bmatrix} 3.832776378145 \\ 3.282956768 \end{bmatrix}.$$

Hence, we compute

$$d^2 = -\nabla F_\mu(x^2) + \alpha_1 d^1, \text{ where } \alpha_1 = \frac{\left\| \nabla F_\mu(x^2) \right\|^2}{\left\| \nabla F_\mu(x^1) \right\|^2} = 0.0083.$$

This gives

$$d^2 = \begin{bmatrix} -3.4463 \\ -3.033957 \end{bmatrix}.$$

The line search problem to minimize $F_\mu(x^2 + \lambda d^2)$ over $\lambda \geq 0$ yields the optimal step length $\lambda^* = 0.0041$. Hence, the next iterate is given by

$$x^3 = x^2 + \lambda^* d^2 = [1.172123914, \ 1.107560776]^t.$$

We can now reset, and continue (with a search along $d^3 = -\nabla F_\mu(x^3)$), until $\left\| \nabla F_\mu \right\|$ is sufficiently small).

9.7 a. The KKT conditions, which are necessary for optimality here, yield
$$\begin{bmatrix} 3x_1^2 \\ 3x_2^2 \end{bmatrix} - v \begin{bmatrix} 1 \\ 1 \end{bmatrix} = \begin{bmatrix} 0 \\ 0 \end{bmatrix}, \quad \text{i.e., } x_1 = \pm\sqrt{v/3} \text{ and } x_2 = \pm\sqrt{v/3}. \text{ But}$$
$x_1 + x_2 = 1 \Rightarrow x_1 = x_2 = 0.5$ and $v = 3/4$ yields the unique KKT solution, and thus provides the optimal solution (since an optimum exists). Hence, $x^* = (0.5, \ 0.5)$.

b. Consider $(x_1, x_2) = (-\gamma\mu, 1)$, where $\gamma > 0$. Then the objective function value of the penalty problem is equal to $-\gamma^3\mu^3 + 1 + \mu^3\gamma^2 = \gamma^2\mu^3(1 - \gamma) + 1$, which decreases without

90

bound as $\gamma \to \infty$ for any $\mu > 0$. Hence, for each $\mu > 0$, the penalty problem is unbounded.

c. Theorem 9.2.2 requires the penalty problem to have an optimum for each $\mu > 0$, and so this result is not applicable for the present problem.

d. The penalty problem with the added bounds is given by

$$\text{Minimize } x_1^3 + x_2^3 + \mu(x_1 + x_2 - 1)^2$$
$$\text{subject to } -1 \le x_1 \le 1, \ -1 \le x_2 \le 1.$$

For convenience, we assume that $\mu > 1$ is large enough. It is readily verified that the KKT solution at which the gradient of the objective vanishes is better than any other solution at which either x_1 or x_2 are at any of their bounds. Since the KKT conditions are necessary for optimality in this case, this KKT point must be the optimal solution. To derive this solution, we examine the system

$$\begin{bmatrix} 3x_1^2 + 2\mu(x_1 + x_2 - 1) \\ 3x_2^2 + 2\mu(x_1 + x_2 - 1) \end{bmatrix} = \begin{bmatrix} 0 \\ 0 \end{bmatrix}.$$

Subtracting the second equation from the first yields $x_1^2 = x_2^2$, i.e., $x_1 = \pm x_2$. But $x_1 = -x_2$ gives an objective value of μ, which is inferior, for example, to the solution $(1/2, 1/2)$ of objective value $1/4$. Hence, $x_1 = x_2$ at optimality, which yields $3x_1^2 + 4\mu x_1 - 2\mu = 0$, or

$$x_1 = x_2 = \frac{-4\mu \pm \sqrt{16\mu^2 + 24\mu}}{6}. \text{ For } \mu \text{ large enough, feasibility}$$

requires the $+\sqrt{\ }$ choice, which yields the optimal solution $x \equiv x_\mu$ given by

$$x_{\mu 1} = x_{\mu 2} = \frac{-2\mu + \sqrt{4\mu^2 + 6\mu}}{3} = \frac{2\mu}{(2\mu + \sqrt{4\mu^2 + 6\mu})} = \frac{1}{1 + \sqrt{1 + \frac{3}{2\mu}}}.$$

Hence, $(x_{\mu 1}, x_{\mu 2}) \to (1/2, 1/2)$ as $\mu \to \infty$, as expected from Part (a) and Theorem 9.2.2.

91

9.8 a. Let (a_i, b_i), $i = 1, \ldots, n$, denote the coordinates of the $m = 4$ existing facilities. Then the problem is as follows:

Minimize $f(x) \equiv \dfrac{1}{2} \sum\limits_{i=1}^{m} \left[(x_1 - a_i)^2 + (x_2 - b_i)^2 \right]$

subject to $3x_1 + 2x_2 = 6$

$(x_1, x_2) \geq 0.$

b. The Hessian of the objective function is given by $\begin{bmatrix} m & 0 \\ 0 & m \end{bmatrix}$, which is PD. Hence, f is strictly convex.

c. Denoting (u_1, u_2, v) as the respective Lagrange multipliers associated with the constraints $x_1 \geq 0$, $x_2 \geq 0$, and $3x_1 + 2x_2 = 6$ (all multiplied through by -1 for writing them in standard form for convenience), we get the following system of KKT conditions, where $m = 4$:

$$mx_1 - \sum_{i=1}^{m} a_i - u_1 - 3v = 0$$

$$mx_2 - \sum_{i=1}^{m} b_i - u_2 - 2v = 0$$

$$3x_1 + 2x_2 = 6$$

$$(x_1, x_2) \geq 0, \; (u_1, u_2) \geq 0$$

$$u_1 x_1 = u_2 x_2 = 0.$$

This system yields the unique solution $u_1 = u_2 = 0$, $(x_1, x_2) = \left(\dfrac{9}{26}, \dfrac{129}{52} \right)$, and $v = \dfrac{19}{13}$. Since the KKT conditions are sufficient (and necessary) in this case, this is an optimal solution.

d. The quadratic penalty function approach yields the following penalty problem:

$$\mathbf{PP}(\mu): \; \underset{(x_1, x_2)}{\text{Minimize}} \left\{ \dfrac{1}{2} \sum_{i=1}^{m} \left[(x_1 - a_i)^2 + (x_2 - b_i)^2 \right] \right.$$

$$\left. + \dfrac{1}{2} \mu \left[(3x_1 + 2x_2 - 6)^2 + \sum_{j=1}^{2} \max^2 \{0, \, -x_j\} \right] \right\},$$

92

where $\mu/2$ is the penalty parameter. Denoting $F_\mu(x)$ as the objective function of Problem PP(μ), we have that

$$\nabla F_\mu(x) = \begin{bmatrix} (4 + 9\mu)x_1 + 6\mu x_2 - \mu \max\{0, -x_1\} - [\sum_i a_i + 18\mu] \\ 6\mu x_1 + 4(1 + \mu)x_2 - \mu \max\{0, -x_2\} - [\sum_i b_i + 12\mu] \end{bmatrix}.$$

We start at $x^1 = (0, 0)$ and apply the Fletcher and Reeves conjugate gradient method using $\mu = 10$.

Iteration 1: We get $\nabla F_\mu(x^1) = \begin{bmatrix} -177 \\ -127 \end{bmatrix}$, and so $d^1 = \nabla F_\mu(x^1) = \begin{bmatrix} 177 \\ 127 \end{bmatrix}$. Performing a line search to Minimize $\underset{\lambda \geq 0}{\nabla F_\mu}(x^1 + \lambda d^1)$, yields λ as a solution to

$$\nabla F_\mu(x^1 + \lambda d^1)^t d^1 = 0, \quad \text{i.e.,} \quad \begin{pmatrix} 24258\lambda - 177 \\ 16208\lambda - 127 \end{pmatrix}^t \begin{bmatrix} 177 \\ 127 \end{bmatrix} = 0, \quad \text{which}$$

yields $\lambda = 0.0074712$, or $(x^1 + \lambda d^1) = \begin{bmatrix} 1.3224 \\ 0.9488 \end{bmatrix} = x^2$.

Iteration 2: We next compute $\nabla F_\mu(x^2) = \begin{bmatrix} 4.2336 \\ -5.9088 \end{bmatrix}$. Hence, we compute

$$d^2 = -\nabla F_\mu(x^2) + \alpha_1 d^1 \text{ where } \alpha_1 = \frac{\left\| \nabla F_\mu(x^2) \right\|^2}{\left\| \nabla F_\mu(x^1) \right\|^2}.$$

This yields $\alpha_1 = 0.0011133$, with

$$d^2 = \begin{bmatrix} 4.2336 \\ 5.9088 \end{bmatrix} + 0.0011133 \begin{bmatrix} 177 \\ 127 \end{bmatrix} = \begin{bmatrix} -4.03655 \\ 6.0502 \end{bmatrix}.$$

Performing a line search along d^2 from x^2, we effectively compute the optimal step length λ as a solution to $\nabla F_\mu(x^2 + \lambda d^2)^t d^2 = 0$, i.e., so long as $x^2 + \lambda d^2 \geq 0$, which yields $\lambda = 0.24975$, thus

93

producing $\quad x^3 = x^2 + \lambda d^2 = \begin{bmatrix} 0.3142 \\ 2.4598 \end{bmatrix}$. $\quad$ This $\quad$ yields

$\nabla F_\mu(x^3) = \begin{bmatrix} 0.1228 \\ 0.0832 \end{bmatrix}$, which is close (albeit with a loose tolerance) to

$\begin{bmatrix} 0 \\ 0 \end{bmatrix}$. (Note that since the optimization essentially involved minimizing a strictly convex quadratic function (since the $\max^2\{0, -x_j\}$-terms did not play a role here), we expect the conjugate gradient method to converge in $n = 2$ iterations. Thus, with exact computations, we should have achieved $\nabla F_\mu(x^3) = \begin{bmatrix} 0 \\ 0 \end{bmatrix}$.)

Checking the feasibility of this solution, we see that $3x_1^3 + 2x_2^3 = 5.8622$. We can thus reset the conjugate gradient algorithm, increase μ to 100, and repeat until we attain near-feasibility to a desired tolerance.

9.12 Following the derivation of KKT Lagrange multipliers at optimality as in Section 9.2, and assuming that the conditions of Theorem 9.2.2 hold with $X = \{x : g_i(x) \le 0 \quad$ for $\quad i = m+1,...,m+M; \quad h_i(x) = 0 \quad$ for $i = \ell+1,...,\ell+L\}$, let x_μ be an optimal solution for the following penalty problem for each $\mu > 0$:

PP: $\quad$ Minimize $\quad f(x) + \mu\alpha(x)$

$\quad\quad$ subject to $\quad g_i(x) \le 0$ for $i = m+1,...,m+M$,

$\quad\quad\quad\quad\quad\quad h_i(x) = 0$ for $i = \ell+1,...,\ell+L$,

where $\alpha(x) = \sum\limits_{i=1}^{m} \phi[g_i(x)] + \sum\limits_{i=1}^{\ell} \psi[h_i(x)]$, and where we assume that:

1. All functions f, g_i, h_i, ϕ, and ψ are continuously differentiable.

2. The functions $\phi(\cdot)$ and $\psi(\cdot)$ satisfy (9.1b), and furthermore, $\phi'(y) \ge 0$ for all y, with $\phi'(y) = 0$ if $y \le 0$.

Hence, since x_μ solves the penalty problem PP for any given $\mu > 0$, by the KKT necessary optimality conditions, we can claim the existence of scalars $y_{\mu i}$, $i = m+1,...,m+M$, and $w_{\mu i}$, $i = \ell+1,...,\ell+L$, such that

94

$$\nabla f(x_\mu) + \mu \nabla \alpha(x_\mu) + \sum_{i=m+1}^{m+M} y_{\mu i} \nabla g_i(x_\mu) + \sum_{i=\ell+1}^{\ell+L} w_{\mu i} \nabla h_i(x_\mu) = 0, \qquad (1)$$

$$y_{\mu i} g_i(x_\mu) = 0 \ \text{ for } \ i = m+1,...,m+M, \qquad (2)$$

$$y_{\mu i} \geq 0 \ \text{ for } \qquad i = m+1,...,m+M. \qquad (3)$$

Moreover, by the definition of $\alpha(x)$, we have

$$\mu \nabla \alpha(x_\mu) = \mu \sum_{i=1}^{m} \phi'[g_i(x_\mu)] \nabla g_i(x_\mu) + \mu \sum_{i=1}^{\ell} \psi'[h_i(x_\mu)] \nabla h_i(x_\mu). \qquad (4)$$

Let $\bar{x}$ be an accumulation point of the generated sequence $\{x_\mu\}$, where, without loss of generality, let $\{x_\mu\} \to \bar{x}$. Also, define the sets I_0 and I_1 as follows:

$$I_0 = \{i \in \{1,...,m\} : g_i(\bar{x}) = 0\},$$

$$I_1 = \{i \in \{m+1,...,m+M\} : g_i(\bar{x}) = 0\}.$$

For μ large enough, we have $g_i(x_\mu) < 0$ for any i that is not in $I_0 \cup I_1$, so that

$$\phi'[g_i(x_\mu)] = 0 \ \text{ if } \ i \in \{1,...,m\} \ \text{ and } \ i \notin I_0,$$

$$y_{\mu i} = 0 \qquad \text{ if } i \in \{m+1,...,m+M\} \ \text{ and } \ i \notin I_1 \ \text{ (from (2))}. \qquad (5)$$

Therefore, for μ large enough, we have from (1), (4), and (5) that

$$\nabla f(x_\mu) + \mu \sum_{i \in I_0} \phi'[g_i(x_\mu)] \nabla g_i(x_\mu) + \mu \sum_{i=1}^{\ell} \psi'[h_i(x_\mu)] \nabla h_i(x_\mu) +$$

$$\sum_{i \in I_1} y_{\mu i} \nabla g_i(x_\mu) + \sum_{i=\ell+1}^{\ell+L} w_{\mu i} \nabla h_i(x_\mu) = 0. \qquad (6)$$

Let

$$u_{\mu i} = \begin{cases} \mu \phi'[g_i(x_\mu)] & \text{if } i \in I_0 \\ y_{\mu i} & \text{if } i \in I_1 \\ 0 & \text{if } i \notin I_0 \cup I_1 \end{cases} \qquad (7a)$$

95

$$v_{\mu i} = \begin{cases} \mu \psi'[h_i(x_\mu)] & \text{for } i = 1,...,\ell \\ w_{\mu i} & \text{for } i = \ell+1,...,\ell+L. \end{cases} \qquad (7b)$$

Notice that $u_{\mu i} \geq 0$ for $i = m+1,...,m+M$ by Assumption 2 and Equation (3). Since $\{x_\mu\} \to \bar{x}$ and all the functions in (7) are continuous, the Lagrange multipliers $\bar{u}$ and $\bar{v}$ at $\bar{x}$ can then be retrieved in the limit as follows:

$$\bar{u}_i = \lim_{\mu \to \infty} u_{\mu i} \text{ for } i = 1,...,m+M,$$
$$\bar{v}_i = \lim_{\mu \to \infty} v_{\mu i} \text{ for } i = 1,...,\ell+L.$$

9.13 First, note that by the definition of $F_E(x)$, we need to show that $\bar{x}$ is a local minimizer for the following problem, denoted by EPP, which equivalently minimizes $F_E(x)$:

EPP: Minimize $f(x) + \mu\left(\sum_{i=1}^{m} y_i + \sum_{i=1}^{\ell} z_i\right)$

subject to $y_i \geq g_i(x)$ for $i = 1,...,m$

$z_i \geq h_i(x)$ and $z_i \geq -h_i(x)$ for $i = 1,...,\ell$

$y_i \geq 0$ for $i = 1,...,m.$

We show that under the assumptions of the problem, the solution $\bar{x}$ satisfies the second-order sufficient conditions for a local minimizer of Problem EPP as given by Theorem 4.4.2.

For this purpose, noting that $g_i(\bar{x}) \leq 0$, $\forall i = 1,...,m$, and $h_i(\bar{x}) = 0$, $\forall i = 1,...,\ell$, since $\bar{x}$ is feasible to the original Problem P, we first show that if

$$\bar{y}_i = \max\{0, g_i(\bar{x})\} = 0 \text{ for } i = 1,...,m$$

$$\bar{z}_i = \max\{h_i(\bar{x}), -h_i(\bar{x})\} = 0 \text{ for } i = 1,...,\ell,$$

then $(\bar{x}, \bar{y}, \bar{z})$ is a KKT point for Problem EPP.

Note that $(\bar{x}, \bar{y}, \bar{z})$ is a KKT point for EPP if there exist scalars u_i^+, u_i^-, v_i^+, and v_i^- associated with the respective constraints $g_i(x) - y_i \leq 0$, $-y_i \leq 0$, $h_i(x) - z_i \leq 0$, and $-h_i(x) - z_i \leq 0$, such that

$$\nabla f(\bar{x}) + \sum_{i=1}^{m} u_i^+ \nabla g_i(\bar{x}) + \sum_{i=1}^{\ell} (v_i^+ - v_i^-)\nabla h_i(\bar{x}) = 0$$

$$\mu - u_i^+ - u_i^- = 0 \ \text{ for } i = 1,...,m$$

$$\mu - v_i^+ - v_i^- = 0 \ \text{ for } i = 1,...,\ell$$

$$u_i^+ \geq 0 \text{ and } u_i^- \geq 0 \ \text{ for } i = 1,...,m \qquad\qquad (1)$$

$$v_i^+ \geq 0 \text{ and } v_i^- \geq 0 \ \text{ for } i = 1,...,\ell$$

$$u_i^+ = 0 \ \text{ for } i \notin I,$$

where $I = \{i \in \{1,...,m\} : g_i(\bar{x}) = 0\}$, and where note that the only constraints of Problem EPP that are not active at $(\bar{x}, \bar{y}, \bar{z})$ are the inequalities $0 = y_i \geq g_i(x)$ for $i \notin I$ (i.e., for which $g_i(\bar{x}) < 0$). Let $\mu \geq$ maximum $\{\bar{u}_i, i \in I, |\bar{v}_i|, i = 1,...,\ell\}$, where $\bar{u}_i$, $i = 1,...,m$, and $\bar{v}_i$, $i = 1,...,\ell$, are Lagrange multipliers for Problem P corresponding to $\bar{x}$. Next, let

$$\bar{u}_i^+ = \bar{u}_i \text{ if } i \in I \text{ and } \bar{u}_i^+ = 0 \text{ if } i \notin I$$

$$\bar{u}_i^- = \mu - \bar{u}_i^+ \text{ for } i = 1,...,m$$

$$\bar{v}_i^+ = \frac{1}{2}(\mu + \bar{v}_i) \text{ for } i = 1,...,\ell$$

$$\bar{v}_i^- = \frac{1}{2}(\mu - \bar{v}_i) \text{ for } i = 1,...,\ell.$$

It can be easily verified that $\bar{u}_i^+$, $\bar{u}_i^-$, $\bar{v}_i^+$ and $\bar{v}_i^-$ satisfy the KKT system (1). Therefore, $(\bar{x}, \bar{y}, \bar{z})$ is a KKT point for Problem EPP.

Next, we need to show that the Hessian matrix of the restricted Lagrange function for Problem EPP is positive definite on the cone C_μ defined below. For notational simplicity and without loss of generality, we consider $\mu \geq$ maximum $\{\bar{u}_i, i \in I, |\bar{v}_i|, i = 1,...,\ell\}$. In this case, the

97

Lagrange multipliers $\bar{u}^-$, v^+, and $\bar{v}^-$ all take on positive values. Now, consider the following sets:

$$I^+ = \{i \in I : \bar{u}_i > 0\} \text{ and } I^0 = \{i \in I : \bar{u}_i = 0\},$$

and notice that $\bar{u}_i^+ > 0$ only if $i \in I^+$ and is zero otherwise.

The cone C_μ for Problem EPP as per Theorem 4.4.2 is the cone of all nonzero vectors $[d_x^t \ d_y^t \ d_z^t]$ such that

$$\nabla g_i(\bar{x})^t d_x - e_i^t d_y = 0 \qquad \text{for } i \in I^+,$$
$$-e_i^t d_y = 0 \qquad \text{for } i = 1,...,m,$$
$$\nabla h_i(\bar{x})^t d_x - e_i^t d_z = 0 \qquad \text{for } i = 1,...,\ell,$$
$$-\nabla h_i(\bar{x})^t d_x - e_i^t d_z = 0 \qquad \text{for } i = 1,...,\ell,$$
$$\nabla g_i(\bar{x})^t d_x - e_i^t d_y \le 0 \qquad \text{for } i \in I^0,$$

where e_i is the ith unit vector. Since $e_i^t d_y = 0$ for $i = 1,...,m$, we necessarily obtain $d_y = 0$, and furthermore, the last and the first of the above conditions reduce to $\nabla g_i(\bar{x})^t d_x = 0$ for $i \in I^+$ and $\nabla g_i(\bar{x})^t d_x \le 0$ for $i \in I^0$, respectively. Moreover, the third and fourth conditions yield $d_z = 0$ (upon summing these), and $\nabla h_i(\bar{x})^t d_x = 0$ for $i = 1,...,\ell$. Hence, the cone C_μ is the cone of all nonzero vectors $[d^t \ 0^t \ 0^t]$ such that

$$\nabla g_i(\bar{x})^t d = 0 \text{ for } i \in I^+, \ \nabla g_i(\bar{x})^t d \le 0 \text{ for } i \in I^0, \text{ and}$$
$$\nabla h_i(\bar{x})^t d = 0 \text{ for } i = 1,...,\ell. \tag{2}$$

Next, consider the Hessian matrix of the restricted Lagrange function $L(x, y, z)$ for Problem EPP, where by (4.25), we have

$$L(x, y, z) = f(x) + \mu\left(\sum_{i=1}^{m} y_i + \sum_{i=1}^{\ell} z_i\right) + \sum_{i \in I} \bar{u}_i^+ (g_i(x) - y_i) - \sum_{i=1}^{m} \bar{u}_i^- y_i$$

98

$$+ \sum_{i=1}^{\ell} \overline{v}_i^+ (h_i(x) - z_i) - \sum_{i=1}^{\ell} \overline{v}_i^- (h_i(x) + z_i).$$

Therefore, the Hessian matrix $\nabla^2 L(\overline{x}, \overline{y}, \overline{z})$ of $L(x, y, z)$ evaluated at $(\overline{x}, \overline{y}, \overline{z})$ is given by

$$\nabla^2 L(\overline{x}, \overline{y}, \overline{z}) = \begin{bmatrix} L_1 & 0 & 0 \\ 0 & 0 & 0 \\ 0 & 0 & 0 \end{bmatrix},$$

where $L_1 = \nabla^2 f(\overline{x}) + \sum_{i \in I} \overline{u}_i^+ \nabla^2 g_i(\overline{x}) + \sum_{i=1}^{\ell} (\overline{v}_i^+ - \overline{v}_i^-) \nabla^2 h_i(\overline{x})$, and the remaining blocks are zero matrices of appropriate sizes.

Now, consider the quadratic form

$$[d_x^t \ d_y^t \ d_z^t] \nabla^2 L(\overline{x}, \overline{y}, \overline{z}) \ [d_x^t \ d_y^t \ d_z^t]^t.$$

Due to the structure of $\nabla^2 L(\overline{x}, \overline{y}, \overline{z})$ we obtain

$$[d_x^t \ d_y^t \ d_z^t] \nabla^2 L(\overline{x}, \overline{y}, \overline{z}) [d_x^t \ d_y^t \ d_z^t]^t = d_x^t L_1 d_x$$

$$= d_x^t [\nabla^2 f(\overline{x}) + \sum_{i \in I} \overline{u}_i^+ \nabla^2 g_i(\overline{x}) + \sum_{i=1}^{\ell} (\overline{v}_i^+ - \overline{v}_i^-) \nabla^2 h_i(\overline{x})] d_x$$

$$= d_x^t [\nabla^2 f(\overline{x}) + \sum_{i \in I} \overline{u}_i \nabla^2 g_i(\overline{x}) + \sum_{i=1}^{\ell} \overline{v}_i \nabla^2 h_i(\overline{x})] d_x = d^t \nabla^2 L(\overline{x}) d,$$

where $L(x)$ is the restricted Lagrange function for Problem P, and $d \equiv d_x$. But the quadratic form $d^t \nabla^2 L(\overline{x}) d$ is known to be positive for all nonzero vectors d such that $\nabla^2 g_i(\overline{x})^t d = 0$ for $i \in I^+$, $\nabla g_i(\overline{x})^t d \leq 0$ for $i \in I^0$, and $\nabla h_i(\overline{x})^t d = 0$ for $i = 1,...,\ell$, since $\overline{x}$ satisfies the second-order sufficient conditions for Problem P as given in Theorem 4.4.2. By Equation (2), we can thus assert that the Hessian of the restricted Lagrange function $L(x, y, z)$ at $(\overline{x}, \overline{y}, \overline{z})$ for Problem EPP is positive definite on the cone C_μ. Therefore, $(\overline{x}, \overline{y}, \overline{z})$ satisfies the conditions of Theorem 4.4.2 for the Problem EPP, and hence is a strict local minimizer for this problem. This means that $\bar{x}$ is a local minimizer of the function F_E, which completes the proof. $\square$

9.14 As in Equation (9.26), let

$$F_{ALAG}(x,\ u,\ v) = f(x) + \mu \sum_{i=1}^{m} \max^2\{g_i(x) + \frac{u_i}{2\mu}, 0\}$$

$$- \sum_{i=1}^{m} \frac{u_i^2}{4\mu} + \sum_{i=1}^{\ell} v_i h_i(x) + \mu \sum_{i=1}^{\ell} h_i^2(x).$$

By assumption, x_k minimizes $F_{ALAG}(\bar{x}, \bar{u}, \bar{v})$, where $\bar{u}$ and $\bar{v}$ are given vectors, so that $\nabla_x F_{ALAG}(x_k, \bar{u}, \bar{v}) = 0$. This means that

$$\nabla f(x_k) + 2\mu \sum_{i=1}^{m} \max\{g_i(x) + \frac{u_i}{2\mu}, 0\} \nabla g_i(x_k)$$

$$+ \sum_{i=1}^{\ell} \bar{v}_i \nabla h_i(x_k) + 2\mu \sum_{i=1}^{\ell} h_i(x_k) \nabla h_i(x_k) = 0,$$

or equivalently,

$$\nabla f(x_k) + 2\mu \sum_{i=1}^{m} \max\{g_i(x) + \frac{u_i}{2\mu}, 0\} \nabla g_i(x_k) +$$

$$\sum_{i=1}^{\ell} (\bar{v}_i + 2\mu h_i(x_k)) \nabla h_i(x_k) = 0. \qquad (1)$$

If $L(x,\ u,\ v) = f(x) + \sum_{i=1}^{m} u_i g_i(x) + \sum_{i=1}^{\ell} v_i h_i(x)$, then

$$\nabla_x L(x_k, \bar{u}_{new}, \bar{v}_{new}) = \nabla f(x_k) + \sum_{i=1}^{m} (\bar{u}_{new})_i \nabla g_i(x_k) + \sum_{i=1}^{\ell} (\bar{v}_{new})_i \nabla h_i(x_k).$$

The requirement that $\nabla_x L(x_k, \bar{u}_{new}, \bar{v}_{new}) = 0$ together with Equation (1) (known to hold), yields the following expressions for $\bar{u}_{new}$ and $\bar{v}_{new}$:

$$(\bar{u}_{new})_i = 2\mu \max\{g_i(x_k) + \frac{\bar{u}_i}{2\mu}, 0\} \text{ for } i = 1,...,m, \qquad (2)$$

$$(\bar{v}_{new})_i = \bar{v}_i + 2\mu h_i(x_k) \text{ for } i = 1,...,\ell. \qquad (3)$$

By taking the factor 2μ within the maximand in (2) and factoring out $\bar{u}_i$, we get

$$(\bar{u}_{new})_i = \bar{u}_i + \max\{2\mu g_i(x_k), -\bar{u}_i\} \text{ for } i = 1,...,m. \tag{4}$$

Equations (4) and (3) show that $\bar{u}_{new}$ and $\bar{v}_{new}$ are as given by Equations (9.27) and (9.19), respectively. $\square$

9.16 We are given that $\bar{x}$ is a KKT point for P, and that $\bar{u}$ and $\bar{v}$ are the corresponding Lagrange multipliers. Therefore,

$$\nabla f(\bar{x}) + \sum_{i=1}^{m} \bar{u}_i \nabla g_i(\bar{x}) + \sum_{i=1}^{\ell} \bar{v}_i \nabla h_i(\bar{x}) = 0$$

$g_i(\bar{x}) \leq 0$ for $i = 1,...,m$

$h_i(\bar{x}) = 0$ for $i = 1,...,\ell$

$\bar{u}_i g_i(\bar{x}) = 0$ and $\bar{u}_i \geq 0$ for $i = 1,...,m$.

We are also given that the second-order sufficiency conditions and strict complementarity hold at $(\bar{x}, \bar{u}, \bar{v})$ (so that $I^+ = I$ in Theorem 4.4.2). Thus, we have

$$d^t \nabla^2 L(\bar{x})d > 0 \text{ for all } d \in C = \{d \neq 0: \nabla g_i(\bar{x})^t d = 0 \text{ for } i \in I,$$

$$\nabla h_i(\bar{x})^t d = 0 \text{ for } i = 1,...,\ell\}.$$

The KKT conditions for Problem P$'$ are:

$$\nabla f(x) + \sum_{i=1}^{m} u_i \nabla g_i(x) + \sum_{i=1}^{\ell} v_i \nabla h_i(x) = 0$$

$$2s_i u_i = 0 \text{ for } i = 1,...,m$$

$$g_i(x) + s_i^2 = 0 \text{ for } i = 1,...,m$$

$$h_i(x) = 0 \text{ for } i = 1,...,\ell.$$

Readily, if $(\bar{x}, \bar{u}, \bar{v})$ is a KKT point for P, then $(\bar{x}, \bar{s}, \bar{u}, \bar{v})$, where $\bar{s}_i^2 = -g_i(\bar{x})$, $i = 1,...,m$, is a KKT point for P'. The Hessian of the restricted Lagrangian function for Problem P' at $(\bar{x}, \bar{s})$ is given by

$$H(\bar{x}, \bar{s}) = \begin{bmatrix} \nabla^2 L(\bar{x}) & 0 \\ 0 & D \end{bmatrix},$$

where D is a diagonal matrix with $2\bar{u}_i$, $i = 1,...,m$, along the diagonal. We need to show that the quadratic form $[d_x^t \ \ d_s^t]H\begin{bmatrix} d_x \\ d_s \end{bmatrix}$ is positive definite on the cone

$$C' = \{ \begin{bmatrix} d_x \\ d_s \end{bmatrix} \neq \begin{bmatrix} 0 \\ 0 \end{bmatrix} : \nabla g_i(\bar{x})^t d_x + 2\bar{s}_i d_{si} = 0 \text{ for } i = 1,...,m,$$

$$\nabla h_i(\bar{x})^t d_x = 0 \text{ for } i = 1,...,\ell\}.$$

Note that by the strict complementarity condition and the definition of s, we have $\bar{s}_i = 0$ and $\bar{u}_i > 0$ for each $i \in I$. Also, $\bar{u}_i = 0$ and $\bar{s}_i > 0$ for $i \notin I$. Therefore,

$$[d_x^t \ \ d_s^t]H\begin{bmatrix} d_x \\ d_s \end{bmatrix} = d_x^t \nabla^2 L(\bar{x})d_x + 2\sum_{i=1}^m d_{si}^2 \bar{u}_i$$

$$= d_x^t \nabla^2 L(\bar{x})d_x + 2\sum_{i \in I} d_{si}^2 \bar{u}_i, \tag{1}$$

and the system

$$\nabla g_i(\bar{x})^t d_x + 2\bar{s}_i d_{si} = 0 \text{ for } i = 1,...,m, \tag{2}$$

is equivalent to

$$\nabla g_i(\bar{x})^t d_x = 0 \text{ for } i \in I, \tag{3a}$$

$$\nabla g_i(\bar{x})^t d_x + 2\bar{s}_i d_{si} = 0 \text{ for } i \notin I. \tag{3b}$$

Hence, for any $\begin{bmatrix} d_x \\ d_s \end{bmatrix} \in C'$, we have from (2) and (3) that if $d_x \neq 0$, then

$d_x \in C$, which implies that $d_x^t \nabla^2 L(\overline{x}) d_x > 0$ in (1) with $2 \sum_{i \in I} d_{si}^2 \overline{u}_i \geq 0$.

On the other hand, if $d_x = 0$, then we have $d_{si} = 0$ for $i \notin I$ from (2)

since $\overline{s}_i > 0$ for $i \notin I$, and so $(d_{si}, i \in I) \neq 0$ since $d_s \neq 0$. Thus, in

this case, we get $2 \sum_{i \in I} d_{si}^2 \overline{u}_i > 0$ in (1) with $d_x^t \nabla^2 L(\overline{x}) d_x = 0$. Hence, in

either case, the expression in (1) is positive-valued. Therefore, $H(\overline{x}, \overline{s})$ is

positive definite on the cone C', which completes the proof. $\square$

If the strict complementarity condition does not hold, then we can only
claim that $H(\overline{x}, \overline{s})$ is positive semidefinite on the cone C' since we

could have $\begin{bmatrix} d_x \\ d_s \end{bmatrix} \in C'$ with $d_x = 0$, $d_s \neq 0$, but yet the right-hand side

of (1) being equal to zero.

9.19 The corresponding barrier problem is given as follows:

Minimize $\quad f(x) + \mu \sum_{i=1}^{m} \phi[g_i(x)]$

subject to $\quad g_i(x) < 0 \;$ for $\; i = 1,...,m,$

$\qquad\qquad g_i(x) \leq 0 \;$ for $\; i = m+1,...,m+M,$

$\qquad\qquad h_i(x) = 0 \;$ for $\; i = 1,...,\ell.$

We assume the following as per Section 9.4:

1. The function ϕ satisfies (9.28) and is continuously differentiable.

2. The functions f, g_i, $i = 1,...,m+M$, and $h_i(x)$, $i = 1,...,\ell$, are
   continuously differentiable.

3. The assumptions of Lemma 9.4.2 and Theorem 9.4.3 hold with

   $X \equiv \{x : g_i(x) \leq 0, \; i = m+1,...,m+M, \; h_i(x) = 0, \; i = 1,...,\ell\}.$

4. The optimal solution $\overline{x}$ to the problem

   $\min\{f(x) : g_i(x) \leq 0, \; i = 1,...,m+M, \; h_i(x) = 0, \; i = 1,...,\ell\}$

103

obtained as an accumulation point of the sequence $\{x_\mu\}$ is a regular point.

For simplicity, and without loss of generality, consider the case when $\{x_\mu\}$ itself converges to $\bar{x}$. Define the following two sets I_0 and I_1:

$$I_0 = \{i \in \{1,...,m\} : g_i(\bar{x}) = 0\}$$
$$I_1 = \{i \in \{m+1,...,m+M\} : g_i(\bar{x}) = 0\}.$$

Since $\bar{x}$ is a regular point, there exist unique Lagrange multipliers $\bar{u}_i$, $i = 1,...,m+M$, and $\bar{v}_i$, $i = 1,...,\ell$, such that

$$\nabla f(\bar{x}) + \sum_{i=1}^{m} \bar{u}_i \nabla g_i(\bar{x}) + \sum_{i=m+1}^{m+M} \bar{u}_i g_i(\bar{x}) + \sum_{i=1}^{\ell} \bar{v}_i \nabla h_i(\bar{x}) = 0 \qquad (1)$$

$$\bar{u}_i \geq 0 \text{ for } i = 1,...,m+M$$

$$\bar{u}_i = 0 \text{ for } i \notin I_0 \cup I_1.$$

By assumption, x_μ solves the barrier problem and hence, for each $\mu > 0$, there exist Lagrange multipliers $y_{\mu i}$, $i = m+1,...,m+M$, and $w_{\mu i}$, $i = 1,...,\ell$, such that

$$\nabla f(x_\mu) + \mu \sum_{i=1}^{m} \phi'\{[g_i(x_\mu)]\nabla g_i(x_\mu)$$

$$+ \sum_{i=m+1}^{m+M} y_{\mu i} \nabla g_i(x_\mu) + \sum_{i=1}^{\ell} w_{\mu i} \nabla h_i(x_\mu) = 0, \qquad (2)$$

$$y_{\mu i} \geq 0 \text{ for } i = m+1,...,m+M,$$

$$y_{\mu i} g_i(x_\mu) = 0 \text{ for } i = m+1,...,m+M.$$

We further assume that $\{y_{\mu i}\}$ for $i = m+1,...,m+M$ and $\{w_{\mu i}\}$ for $i = 1,...,\ell$ are contained within a compact set, and hence have convergent subsequences. Without loss of generality, suppose that $\{y_{\mu i}\} \to y_i$ for $i = m+1,...,m+M$, and that $\{w_{\mu i}\} \to w_i$ for $i = 1,...,\ell$.

By the above set of assumptions, since $x_\mu \to \bar{x}$ as $\mu \to 0^+$, we obtain

$$\{\nabla f(x_\mu)\} \to \nabla f(\bar{x}),$$

$$\{\nabla g_i(x_\mu)\} \to \nabla g_i(\bar{x}) \text{ for } i = 1,...,m + M,$$

$$\{\nabla h_i(x_\mu)\} \to \nabla h_i(\bar{x}) \text{ for } i = 1,...,\ell,$$

$$\mu\phi'[g_i(x_\mu)] \to \text{ some } y_i \text{ for } i = 1,...,m, \text{ where } y_i = 0 \text{ if } i \notin I_0$$

$$\{y_{\mu i}\} \to y_i \geq 0 \text{ for } i = m + 1,...,m + M, \text{ where } y_i = 0 \text{ if } i \notin I_1.$$

$$\{w_{\mu i}\} \to w_i \text{ for } i = 1,...,\ell.$$

In particular, we get from (2) in the limit as $\mu \to 0^+$ that

$$\nabla f(\bar{x}) + \sum_{i=1}^{m+M} y_i \nabla g_i(\bar{x}) + \sum_{i=1}^{\ell} w_i \nabla h_i(\bar{x}) = 0, \tag{3}$$

where $y_i = 0$ for $i \notin I_0 \cup I_1$. Since $\bar{x}$ is a regular point, by the uniqueness of the multipliers satisfying Equation (1), we have from (1) and (3) that the unique Lagrange multipliers $\bar{u}_i$, $i = 1,...,m + M$, and $\bar{v}_i$, $i = 1,...,\ell$, can be derived as limits of $\{y_{\mu i}\}$ and $\{w_{\mu i}\}$, respectively. That is

$$\bar{u}_i = y_i = \lim_{\mu \to 0^+} \mu\phi'[g_i(x_\mu)] \text{ for } i = 1,...,m,$$

$$\bar{u}_i = y_i = \lim_{\mu \to 0^+} y_{\mu i} \text{ for } i = m + 1,...,m + M,$$

$$\bar{v}_i = \lim_{\mu \to 0^+} w_{\mu i} \text{ for } i = 1,...,\ell.$$

9.32 For the problem in Example 9.51, where $A = [1 \quad 2]$ and $b = 2$, using Equation (9.43), we obtain the following closed-form expressions for the direction vectors d_v, d_u, and d_x at $\bar{v}$, $\bar{u}$, and $\bar{x}$, respectively:

$$d_v = \frac{2\bar{u}_1\bar{u}_2 - \hat{\mu}(2\bar{u}_1 + \bar{u}_2)}{\bar{x}_1\bar{u}_2 + 4\bar{x}_2\bar{u}_1} \tag{1a}$$

$$d_u = -\begin{bmatrix} 1 \\ 2 \end{bmatrix} d_v \tag{1b}$$

$$d_x = \left[\frac{\hat{\mu} - \bar{x}_1\bar{u}_1 + \bar{x}_1 d_v}{\bar{u}_1}, \frac{\hat{\mu} - \bar{x}_2\bar{u}_2 + 2\bar{x}_2 d_v}{\bar{u}_2} \right]^t. \tag{1c}$$

Consequently, given any iterate $\bar{w} = (\bar{x}, \bar{u}, \bar{v})$ and given $\bar{\mu}$, we set $\hat{\mu} = \beta\bar{\mu}$ (where $\beta = 1 - \dfrac{0.35}{\sqrt{2}} = 0.7525127$), and we compute (d_x, d_u, d_v) from (1) and then obtain the next iterate $\hat{w} = (\hat{x}, \hat{u}, \hat{v})$ according to

$$\hat{x} = \bar{x} + d_x, \quad \hat{u} = \bar{u} + d_u, \quad \text{and} \quad \hat{v} = \bar{v} + d_v. \tag{2}$$

We then continue until the duality gap given by

$$3\bar{x}_1 - \bar{x}_2 - 2\bar{v} = n\bar{\mu} < \varepsilon \text{ for some tolerance } \varepsilon > 0. \tag{3}$$

If the starting vectors are $x_0 = [2/9, 8/9]^t$, $u_0 = [4, 1]^t$, $\mu_0 = 8/9$, and $v_0 = -1$, then the primal-dual path following algorithm produces the following results:

Iteration 1: (with d_v, d_u, and d_x computed as in Example 9.5.1):

$\mu_1 = \beta\mu_0 = 0.6689001$

$x_1 = \hat{x} = [0.17484, 0.9125797]^t$, $u_1 = \hat{u} = [3.8629302, 0.7258604]^t$,

$v_1 = \hat{v} = -0.8629302$. The duality gap equals $2\mu_1 = 1.3378002$.

Iteration 2:

$\mu_2 = \beta\mu_1 = 0.5033558$

Using Equation (1), we get

$$d_v = 0.0951426, \quad d_u = \begin{bmatrix} -0.0951426 \\ -0.1902852 \end{bmatrix}, \quad \text{and} \quad d_x = \begin{bmatrix} -0.0402296 \\ 0.0201151 \end{bmatrix}.$$

This yields

$$x_2 = [0.1346104, 0.9326948]^t$$

$$u_2 = [3.7677876, 0.5355752]^t \text{ and}$$

$$v_2 = -0.7677876.$$

106

Note that the duality gap is now given by

$$[3, \ -1]^t x_2 - 2v_2 = 2\mu_2 = 1.0067116.$$

The iterations can now continue until the duality gap is sufficiently small. (With $\varepsilon = 0.001$, we have that after 25 more iterations, i.e., after obtaining x_{27}, the duality gap will equal $0.0008233 < \varepsilon = 0.001$.)

CHAPTER 10:

METHODS OF FEASIBLE DIRECTIONS

10.3 a. Let $S = \{x : Ax = b,\ x \geq 0\}$. Given $\bar{x} \in S$, let $I = \{j : \bar{x}_j = 0\}$. Then the set of feasible directions at $\bar{x}$ is given by

$$D = \{d : Ad = 0,\ d_j \geq 0 \text{ for } j \in I\}.$$

b. Let $S = \{x : Ax \leq b,\ Qx = q,\ x \geq 0\}$. Given $\bar{x} \in S$, let $I_1 = \{i : A_i\bar{x} = b_i\}$, and $I_2 = \{j : \bar{x}_j = 0\}$, where $A_i x \leq b_i$, $i = 1,\dots,m$, denote the rows of $Ax \leq b$. Then the set of feasible directions at $\bar{x}$ is given by

$$D = \{d : A_i d \leq 0,\ \forall i \in I_1;\ Qd = 0;\ d_j \geq 0,\ \forall j \in I_2\}.$$

c. Let $S = \{x : Ax \geq b,\ x \geq 0\}$. Given $\bar{x} \in S$, and following the same notation as in Part (b) above, let $I_1 = \{i : A_i\bar{x} = b_i\}$ and $I_2 = \{j : \bar{x}_j = 0\}$. Then the set of feasible directions at $\bar{x}$ is given by

$$D = \{d : A_i d \geq 0,\ \forall i \in I_1;\ d_j \geq 0,\ \forall j \in I_2\}.$$

10.4 a. The direction-finding problem in this case decomposes into n (independent) linear one-dimensional problems, one for each d_j, where in the jth problem, the function $\nabla_j d_j$ is minimized over the interval

$$
\begin{aligned}
&[-1,\ 1] &&\text{if } a_j < x_j < b_j \\
&[0,\ 1] &&\text{if } x_j = a_j \\
&[-1,\ 0] &&\text{if } x_j = b_j.
\end{aligned}
$$

Therefore, an optimal solution is given by:

$$
\begin{aligned}
d_j &= -1 &&\text{if } \nabla_j \geq 0 \text{ and } x_j > a_j \\
d_j &= 1 &&\text{if } \nabla_j < 0 \text{ and } x_j < b_j
\end{aligned}
$$

$d_j = 0$ otherwise.

b. Assume that $\nabla_j \neq 0$ for at least one j (else $d = 0$ is optimal). The direction-finding problem is given by:

$$\text{Minimize} \quad \sum_{j=1}^{n} \nabla_j d_j$$

subject to

$$d_j \leq 0 \text{ for } j \text{ such that } x_j = b_j \tag{1}$$

$$d_j \geq 0 \text{ for } j \text{ such that } x_j = a_j \tag{2}$$

$$\sum_{j=1}^{n} d_j^2 \leq 1. \tag{3}$$

It can be easily verified that the vector $d^* \equiv [d_j^*, \ j = 1,...,n]$, where

$$d_j^* = \begin{cases} -\nabla_j \Big/ \left(\sum_{j \in I} \nabla_j^2 \right)^{1/2} & \text{if } j \in I \\ 0 & \text{if } j \notin I, \end{cases}$$

and where $I = \{j : x_j > a_j \text{ and } \nabla_j \geq 0, \text{ or } x_j < b_j \text{ and } \nabla_j < 0\}$ is a feasible solution to this problem. In order to prove that d^* is an optimal solution we first show that the vector d^* is a KKT point for this problem. For this purpose, let u_j, for j such that $x_j = b_j$ or $x_j = a_j$, and w denote Lagrange multipliers associated with the constraints (1) – (3), respectively. The KKT system, aside from the primal feasibility restrictions (already verified above), is given by:

$$\nabla_j + u_j + 2wd_j = 0 \text{ for } j \text{ such that } x_j = b_j$$

$$\nabla_j - u_j + 2wd_j = 0 \text{ for } j \text{ such that } x_j = a_j$$

$$\nabla_j + 2wd_j = 0 \text{ for } j \text{ such that } a_j < x_j < b_j$$

$$u_j d_j = 0 \text{ and } u_j \geq 0 \text{ for all } j \text{ such that } x_j = a_j \text{ or } x_j = b_j$$

$$w \left(\sum_{j=1}^{n} d_j^2 - 1 \right) = 0, \ w \geq 0.$$

It can be easily verified that if $d_j = d_j^*$, $j,...,n$, then

$$u_j = 0 \text{ if } j \in I$$
$$u_j = -\nabla_j \text{ if } j \notin I \text{ and } \nabla_j < 0$$
$$u_j = \nabla_j \text{ if } j \notin I \text{ and } \nabla_j \geq 0$$

$$w = \frac{1}{2}\left(\sum_{j \in I} \nabla_j^2\right)^{1/2}$$

satisfies the KKT system for this problem. Therefore, d^* is a KKT point for the direction-finding problem. Furthermore, the objective function in this problem is linear and all the constraint functions are convex. Therefore, by Theorem 4.3.8, we can assert that d^* is an optimal solution. $\square$

c. Note that $\nabla f(x) = \begin{bmatrix} 6x_1 - 2x_2 - 4 \\ -2x_1 + 8x_2 - 3 \end{bmatrix}$, and the Hessian is

$H = \begin{bmatrix} 6 & -2 \\ -2 & 8 \end{bmatrix}$, which is positive definite. Hence, f is strictly convex. In the computations below, for notational convenience, all vectors are given as row vectors. The method in Part (a) produces the following results:

Iteration 1.

$x^1 = [-2 \ -3]$, $\nabla f(x^1) = [-10 \ -23]$, $f(x^1) = 53$
$d^1 = [1 \ 1]$, $\lambda_1 = \min\{\frac{33}{10}, 2\} = 2$.

Iteration 2.

$x^2 = [0 \ -1]$, $\nabla f(x^2) = [-2 \ -11]$, $f(x^2) = 7$
$d^2 = [0 \ 1]$, $\lambda_2 = \min\{\frac{11}{8}, 2\} = 11/8$.

Iteration 3.

$x^3 = [0 \ 3/8]$, $\nabla f(x^3) = [-19/4 \ 0]$, $f(x^3) = -9/16$,

$d^3 = [0 \ -1]$.

The inner product of $\nabla f(x^3)$ and d^3 is zero and so we stop. The vector $[0 \ 3/8]$ is a KKT point for this problem. Since the assumptions of Theorem 4.3.8 hold, we can conclude that this is an optimal solution.

The method of Part (b) produces the following results:

Iteration 1.

$$x^1 = [-2 \ -3], \ \nabla f(x^1) = [-10 \ -23], \ f(x^1) = 53, \ I = \{1, \ 2\}$$

$$d^1 = \frac{1}{\sqrt{629}}[10 \ 23], \ \lambda_1 = \min\{\frac{629\sqrt{629}}{3912}, \ 2\} = 2.$$

Iteration 2.

$$x^2 = [0 \ -1], \ \nabla f(x^2) = [-2 \ -11], \ f(x^2) = 7, \ I = \{2\}$$

$$d^2 = [0 \ 1], \ \lambda_2 = \min\{\frac{11}{8}, \ 2\} = 11/8.$$

Iteration 3.

$$x^3 = [0 \ 3/8], \ \nabla f(x^3) = [-19/4 \ 0], \ f(x^3) = -9/16, \ I = \{2\}$$

$$d^3 = [0 \ 0].$$

Hence, as above, we terminate with $[0 \ 3/8]$ as an optimal solution. Note that for this problem, the two methods happened to produce the same sequence of iterates.

d. Consider the problem in which the function $f(x_1, \ x_2) = x_1 - 2x_2$ is minimized over the rectangle given in Part (c). Hence, $\nabla f(x) = [1 \ -2]$. Let $x^k = [0 \ 1 - 1/k], \ k = 1,2,...$ For each iterate x^k, the direction-finding map $D(x)$ defined in Part (a) gives $D(x^k) = d^k = [-1 \ 1]$. Thus, we have $\{x^k, \ d^k\} \to (\bar{x}, \ \bar{d})$ where $\bar{x} = [0 \ 1]$ and $\bar{d} = [-1 \ 1]$. However, $D(\bar{x}) = [-1 \ 0] \neq [-1 \ 1]$. This means that $D(x)$ is not closed at $\bar{x}$.

For the direction-finding map $D(x)$ described in Part (b), we obtain using this same example that

$$D(x^k) = d^k = \frac{1}{\sqrt{5}}[-1\ 2], \ \{x^k, \ d^k\} \rightarrow (\bar{x}, \ \bar{d}), \ \text{where} \ \bar{x} = [0\ 1]$$

and $\bar{d} = \frac{1}{\sqrt{5}}[-1\ 2]$, but $D(\bar{x}) = [-1\ 0] \neq \bar{d}$. Hence, in this case as well, the direction-finding map is not closed.

e. For example, for Part (b), see the counterexample given in Example 10.2.3.

10.9 Let P denote the original problem and let DF denote the given direction-finding problem. First of all, note that $d = 0$ with objective value equal to zero is a feasible solution to Problem DF and that the feasible region of Problem DF is bounded since $\|d\|_\infty \leq 1$. Hence, Problem DF has an optimum with a nonpositive optimal objective value. Furthermore, d is an improving feasible direction at x if and only if

$$\nabla f(x)^t d < 0 \ \text{and} \ \nabla g_i(x)^t d \leq 0, \ \forall i \in I, \tag{1}$$

because $\nabla f(x)^t d < 0 \Rightarrow f(x + \lambda d) < f(x), \ \forall 0 \leq \lambda \leq \delta,$ for some $\delta > 0$, by the differentiability of f, and because for each $i \in I$, we have that

$$\nabla g_i(x)^t d \leq 0 \Rightarrow g_i(x + \lambda d) \leq g_i(x), \ \forall \lambda \geq 0,$$

by the pseudoconcavity of g at x.

Consequently, if the optimal objective value of Problem DF (denoted v^*) is zero, then there does not exist any improving feasible direction at x, for else, (1) would have a solution, which by scaling d so that $\|d\|_\infty \leq 1$, would yield a feasible solution to DF with a negative objective value, thus contradicting that $v^* = 0$. On the other hand, if $v^* < 0$, then the optimal solution d^* to Problem DF satisfies (1) and hence yields an improving feasible direction.

10.12 a. If $\hat{x}$ is a Fritz John point for the problem: $\min\{f(x) : g_i(x) \leq 0, \ i = 1,...,m\}$, then there exists a vector $(u_i, \ i \in I)$, where $I = \{i : g_i(\hat{x}) = 0\}$, and a scalar u_0 such that

112

$$\nabla f(\hat{x}) u_0 + \sum_{i \in I} u_i \nabla g_i(\hat{x}) = 0$$

$$u_0 \geq 0, \; u_i \geq 0$$

$(u_0, \; u_i \text{ for } i \in I) \neq (0, \; 0).$

This means that the system $A^t y = 0$, $y \geq 0$, $y \neq 0$, where the columns of A^t are $\nabla f(\hat{x})$ and $\nabla g_i(\hat{x})$, $i \in I$, has a solution (here $y \equiv [u_0, \; u_i \text{ for } i \in I]^t$). Therefore, by Gordan's Theorem, the system $Ad < 0$ has no solution. That is, no (nonzero) vector d exists such that $\nabla f(\hat{x})^t d < 0$ and $\nabla g_i(\hat{x})^t d < 0$, $i \in I$. This implies that the feasible set in the problem:

Minimize z

subject to
$$\nabla f(\hat{x})^t d \leq z$$

$$\nabla g_i(\hat{x})^t d \leq z \text{ for } i \in I,$$

$$d_j \geq -1 \text{ if } \frac{\partial f(\hat{x})}{\partial x_j} > 0$$

$$d_j \leq 1 \text{ if } \frac{\partial f(\hat{x})}{\partial x_j} < 0$$

has no point for which $z < 0$. However, $(\hat{z}, \; \hat{d}) = (0, \; 0)$ is a feasible solution for this problem, and therefore, it must be optimal.

A similar argument can be used to show that if $(\hat{z}, \; \hat{d}) = (0, \; 0)$ solves the foregoing problem, then $\hat{x}$ is a Fritz John point for the problem: $\min\{f(x) : g_i(x) \leq 0, \; i = 1,...,m\}$, noting that the restrictions on the d_j-components are simply used to bound the objective value from below in case there exists a d such that $\nabla f(\hat{x})^t d < 0$ and $\nabla g_i(\hat{x})^t d < 0$, $\forall i \in I$.

b. By the problem formulation, we have $\hat{z} = \max\{\nabla f(\hat{x})^t \hat{d}, \; \nabla g_i(\hat{x})^t \hat{d}$ for $i \in I\}$. Since $\hat{z} < 0$, we necessarily have $\nabla f(\hat{x})^t \hat{d} < 0$. Therefore, $\hat{d}$ is an improving direction of $f(x)$ at $\hat{x}$. Furthermore, since $\nabla g_i(\hat{x})^t \hat{d} < 0$, $\forall i \in I$, there exists $\hat{\lambda} > 0$ such that

113

$g_i(\hat{x} + \lambda \hat{d}) \le 0$ for any $\lambda \in (0, \hat{\lambda})$, $\forall i = 1,...,m$, and so $\hat{d}$ is also a feasible direction at $\hat{x}$. This shows that $\hat{d}$ is an improving feasible direction at $\hat{x}$. □

c. We could likewise select any $k \in I$, and use the following bounding constraints:

$$d_j \ge -1 \text{ if } \frac{\partial g_k(\hat{x})}{\partial x_j} > 0$$

$$d_j \le 1 \text{ if } \frac{\partial g_k(\hat{x})}{\partial x_j} < 0.$$

10.19 Let $Q(d; \bar{x}) = \nabla f(\bar{x})^t + \frac{1}{2} d^t H(\bar{x}) d$.

a. The required second-order approximation problem (QA), is given as follows:

QA: Minimize $Q(d; \bar{x})$
subject to $Ad = 0$
$$d_j \ge 0, \ \forall j \in J_0$$
$$-1 \le d_j \le 1, \ \forall j = 1,...,n.$$

Let d^* denote an optimal solution to Problem QA.

b. Note that Problem QA involves minimizing a strictly convex function over a nonempty polyhedron. Therefore, the KKT conditions are both necessary and sufficient for optimality. If $d^* = 0$, then there exist scalars $u_j \ge 0$, $j \in J_0$, and a vector v such that

$$\nabla Q(0; \bar{x}) - \sum_{j \in J_0} u_j e_j + A^t v = 0,$$

where e_j is the jth unit vector in R^n.

Notice that $\nabla Q(d; \bar{x}) = \nabla f(\bar{x}) + H(\bar{x})d$, which yields $\nabla Q(0; \bar{x})$ $= \nabla f(\bar{x})$. Therefore, if $d^* = 0$ solves the problem QA, then there exist scalars $u_j \geq 0$, $j \in J_0$, and a vector v such that

$$\nabla f(\bar{x}) - \sum_{j \in J_0} u_j e_j + A^t v = 0.$$

This implies that $\bar{x}$ is a KKT point for the original problem.

c. Suppose that $d^* \neq 0$. Then the optimal objective value of Problem QA is given by $Q(d^*; \bar{x}) < 0$, since $d = 0$ is a feasible solution. Therefore, since $\nabla f(\bar{x})^t d^* + \frac{1}{2} d^{*t} H(\bar{x}) d^* < 0$ and $H(\bar{x})$ is a positive definite matrix, we necessarily have $\nabla f(\bar{x})^t d^* < 0$. Thus, d^* is an improving feasible direction for Problem P at $\bar{x}$, where the feasibility of d^* follows directly from the formulation of the constraints of Problem QA. $\square$

10.20 The KKT conditions are given as follows:

$$4x_1 - 2x_2 + 4x_1 u_1 + u_2 - u_3 = 4$$
$$-2x_1 + 4x_2 - u_1 + 5u_2 - u_4 = 6$$
$$u_1(2x_1^2 - x_2) = 0, \; u_2(x_1 + 5x_2 - 5) = 0$$
$$u_3 x_1 = 0, \; u_4 x_2 = 0$$
$$x_1 \geq 0, \; x_2 \geq 0, \; u_i \geq 0 \text{ for } i = 1, 2, 3, 4.$$

The above KKT system has the following unique solution:

$$\bar{x}_1 = (\sqrt{201} - 1)/20$$
$$\bar{x}_2 = (101 - \sqrt{201})/100$$
$$\bar{u}_1 = 0.82243058, \; \bar{u}_2 = 0.93345463, \text{ and } \bar{u}_3 = \bar{u}_4 = 0.$$

By Theorem 9.3.1, a suitable value for μ is any real number such that $\mu \geq 0.93345463$. The Hessian H of the objective function $f(x)$ is given by $H = \begin{bmatrix} 4 & -2 \\ -2 & 4 \end{bmatrix}$. Its eigenvalues are $\lambda_1 = 6$ and $\lambda_2 = 2$. The vectors

$[-1\ 1]^t$ and $[1\ 1]^t$ are eigenvectors corresponding to λ_1 and λ_2, respectively. The unconstrained minimum of $f(x)$ can be found by solving the system $\nabla f(x) = 0$, which yields

$$4x_1 - 2x_2 = 4$$
$$-2x_1 + 4x_2 = 6.$$

The unique solution to the above system is $x_1^* = 7/3$ and $x_2^* = 8/3$.

Hence, the contours of $f(x)$ are elliptical, centered at $(7/3,\ 8/3)^t$, with major and minor axes oriented along $(1,\ 1)^t$ and $(-1,\ 1)^t$, respectively, as depicted in Figure 10.13a.

Successive iterations (beyond Iteration 1 as presented in Example 10.32) of applying Algorithm PSLP to this problem are as follows (note that constant terms $f(x_k) - \nabla f(x_k)^t x_k$ are not included in the objective function of Problem $LP(x_k,\ \nabla_k)$):

Iteration 2.

$x_2 = (0.5,\ 0.9)^t$, $\Delta_2 = (1,\ 1)^t$,

$LP(x_2,\ \Delta_2)$: Minimize $-3.8x_1 - 3.4x_2 + 10\max\{0,\ -0.5 + 2x_1 - x_2\}$
subject to $\quad x_1 + 5x_2 \leq 5$
$\quad 0 \leq x_1 \leq 1.5,\ 0 \leq x_2 \leq 1.9.$

An optimal solution to $LP(x_2,\ \Delta_2)$ is given by $x_1 = 15/22$ and $x_2 = 19/22$, where $\Delta F_{E_2} = -0.13868$ and $\Delta F_{EL_2} = 0.5672$, and so $R_2 < 0$. Therefore, we need to shrink Δ_2 to $(0.5, 0.5)$ and repeat this step.

$LP(x_2,\ \Delta_2)$: Minimize $-3.8x_1 - 3.4x_2 + 10\max\{0,\ -0.5 + 2x_1 - x_2\}$
subject to $\quad x_1 + 5x_2 \leq 5$
$\quad 0 \leq x_1 \leq 1,\ 0.4 \leq x_2 \leq 1.4.$

An optimal solution to $LP(\mathbf{x}_2, \boldsymbol{\Delta}_2)$ is given by $x_1 = 15/22$ and $x_2 = 19/22$, where $\Delta F_{E_2} = -0.13868$ and $\Delta F_{EL_2} = 0.5672$, and so $R_2 < 0$. Therefore, again, we need to shrink $\boldsymbol{\Delta}_2$ to $(0.25, 0.25)$ and repeat this step.

$LP(\mathbf{x}_2, \boldsymbol{\Delta}_2)$: Minimize $-3.8x_1 - 3.4x_2 + 10\max\{0, -0.5 + 2x_1 - x_2\}$
subject to $\quad x_1 + 5x_2 \le 5$
$$0.25 \le x_1 \le 0.75, \ 0.65 \le x_2 \le 1.15.$$

An optimal solution to $LP(\mathbf{x}_2, \boldsymbol{\Delta}_2)$ is given by $x_1 = 15/22$ and $x_2 = 19/22$, where $\Delta F_{E_2} = -0.13868$ and $\Delta F_{EL_2} = 0.5672$, and so $R_2 < 0$. Once again, we need to shrink $\boldsymbol{\Delta}_2$ to $(0.125, 0.125)$ and repeat this step.

$LP(\mathbf{x}_2, \boldsymbol{\Delta}_2)$: Minimize $-3.8x_1 - 3.4x_2 + 10\max\{0, -0.5 + 2x_1 - x_2\}$
subject to $\quad x_1 + 5x_2 \le 5$
$$0.375 \le x_1 \le 0.625, \ 0.785 \le x_2 \le 1.025.$$

An optimal solution to $LP(\mathbf{x}_2, \boldsymbol{\Delta}_2)$ is given by $x_1 = 0.625$ and $x_2 = 0.875$, where $\Delta F_{E_2} = 0.51325$ and $\Delta F_{EL_2} = 0.39$, and so $R_2 = 1.1316 > \rho_2$. We thus accept the new solution $\mathbf{x}_3 = (0.625, 0.875)$ and so amplify the trust region by a factor of 2.

Iteration 3.

$\mathbf{x}_3 = (0.625, 0.875)$, $\boldsymbol{\Delta}_3 = (0.25, 0.25)$.

$LP(\mathbf{x}_3, \boldsymbol{\Delta}_3)$: Minimize $-3.25x_1 - 3.75x_2 + 10\max\{0, -0.78125 + 2.5x_1 - x_2\}$
subject to $\quad x_1 + 5x_2 \le 5$
$$0.375 \le x_1 \le 0.875, \ 0.625 \le x_2 \le 1.125.$$

An optimal solution to $LP(\mathbf{x}_3, \boldsymbol{\Delta}_3)$ is given by $x_1 = 0.6597$ and $x_2 = 0.86806$, where $\Delta F_{E_3} = 0.060282056$ and $\Delta F_{EL_3} = 0.8675$, and so $R_2 = 0.6949$. We thus accept the new iterate as $\mathbf{x}_4$, but retain $\boldsymbol{\Delta}_3$ as the value of $\boldsymbol{\Delta}_4$.

Iteration 4.

$$\mathbf{x}_4 = (0.6597, 0.86806), \ \Delta_4 = (0.25, 0.25).$$

$LP(\mathbf{x}_4, \Delta_4)$:

Minimize
$$-3.09732x_1 - 3.84716x_2 + 10\max\{0, -0.87040818 + 2.6388x_1 - x_2\}$$
subject to $x_1 + 5x_2 \leq 5$
$$0.4097 \leq x_1 \leq 0.9097, \ 0.61806 \leq x_2 \leq 1.11806.$$

An optimal solution to $LP(\mathbf{x}_4, \Delta_4)$ is given by $x_1 = 0.65887$ and $x_2 = 0.86823$, where $\Delta F_{E_4} = 0.021547944$ and $\Delta F_{EL_4} = 0.022137689$, and so $R_2 = 0.97336$. We therefore accept the new iterate as $\mathbf{x}_5$, and we amplify Δ_4 to get Δ_5.

Iteration 5.

$$\mathbf{x}_5 = (0.65887, 0.86823), \ \Delta_5 = (0.5, 0.5).$$

$LP(\mathbf{x}_5, \Delta_5)$:
Minimize
$$-3.10098x_1 - 3.84482x_2 + 10\max\{0, -0.868219353 + 2.63548x_1 - x_2\}$$
subject to $x_1 + 5x_2 \leq 5$
$$0.15887 \leq x_1 \leq 1.15887, \ 0.36823 \leq x_2 \leq 1.366823.$$

An optimal solution to $LP(\mathbf{x}_5, \Delta_5)$ is given by $x_1 = 0.65887$ and $x_2 = 0.86826$, where $\Delta F_{E_5} = 0.0$ and $\Delta F_{EL_5} = 0.0001153$. Since ΔF_{EL_5} is small enough (the direction $\mathbf{d}_5 = (0, 0.00003)^t$ is sufficiently small in norm), we stop; the solution $(x_1, x_2) = (0.65887, 0.86826)$ is close enough to the desired KKT point.

10.25 Algorithm RSQP applied to Example 10.4.3:

Using superscripts for distinguishing vector iterates from its components, we have for the next iteration:

$x^2 = (1.1290322, \ 0.7741936)^t$

$u^2 = (0, \ 1.032258, \ 0, \ 0)^t$

$$\nabla f(x) = \begin{bmatrix} 4x_1 - 2x_2 - 4 \\ 4x_2 - 2x_1 - 6 \end{bmatrix}, \quad \nabla^2 f(x) = \begin{bmatrix} 4 & -2 \\ -2 & 4 \end{bmatrix}$$

$$\nabla f(x^2) = \begin{bmatrix} -1.0322584 \\ -5.16129 \end{bmatrix}$$

$$\nabla^2 L(x^2) = \nabla^2 f(x^2) + u_2 \nabla^2 g_2(x^2) = \nabla^2 f(x^2) = \begin{bmatrix} 4 & -2 \\ -2 & 4 \end{bmatrix}$$

$$g_1(x^2) = 1.7752338, \quad \nabla g_1(x^2) = \begin{bmatrix} 4.5161288 \\ -1 \end{bmatrix}$$

$$g_2(x^2) = 0, \quad \nabla g_2(x^2) = \begin{bmatrix} 1 \\ 5 \end{bmatrix}$$

$$g_3(x^2) = -1.1290322, \quad \nabla g_3(x^2) = \begin{bmatrix} -1 \\ 0 \end{bmatrix}$$

$$g_4(x^2) = -0.7741936, \quad \nabla g_4(x^2) = \begin{bmatrix} 0 \\ -1 \end{bmatrix}.$$

Hence, the direction-finding quadratic program (QP) is given as follows:

QP(x^2, u^2) :

Minimize $\quad -1.0322584d_1 - 5.16129d_2 + \dfrac{1}{2}\left[4d_1^2 + 4d_2^2 - 4d_1d_2\right]$

subject to $\quad 1.7752338 + 4.5161288d_1 - d_2 \leq 0$

$$d_1 + 5d_2 \leq 0$$

$$-1.1290322 - d_1 \leq 0$$

$$-0.7741936 - d_2 \leq 0.$$

At optimality for QP(x^2, u^2), the first two constraints are active, thus yielding the following optimal solution via the KKT conditions:

$d^2 = [-0.411302, \ 0.0822604]^t$

$u^3 = [0.4325705, \ 0.8884432, \ 0, \ 0]^t.$

Taking a unit step along d^2 yields

$$x^3 = x^2 + d^2 = [0.7177302, \ 0.8564454]^t.$$

Note from Figure 10.13(a) that an optimal solution to the given original problem is

$$x^* = [0.6588722, \ 0.8682256]^t.$$

Hence, we are approaching x^* and the iterations can now continue as above.

Algorithm MSQP applied to Example 10.4.3.

In this case, the solution x^2 after the first iteration is given as follows

$$x^2 = [0.6588722, \ 0.8682256]^t.$$

Note that this coincides with x^* above, and its optimality is verified below.

$$u^2 = [0, \ 1.032258, \ 0, \ 0]^t.$$

$$\nabla f(x^2) = \begin{bmatrix} -3.1009624 \\ -3.844842 \end{bmatrix}, \quad \nabla^2 L(x^2) = \begin{bmatrix} 4 & -2 \\ -2 & 4 \end{bmatrix} \ \text{(as before)}$$

$$g_1(x^2) = 0, \ \nabla g_1(x^2) = \begin{bmatrix} 2.6354888 \\ -1 \end{bmatrix}$$

$$g_2(x^2) = 0, \ \nabla g_2(x^2) = \begin{bmatrix} 1 \\ 5 \end{bmatrix}$$

$$g_3(x^2) = -0.6588722, \ \nabla g_3(x^2) = \begin{bmatrix} -1 \\ 0 \end{bmatrix}$$

$$g_4(x^2) = -0.8682256, \ \nabla g_4(x^2) = \begin{bmatrix} 0 \\ -1 \end{bmatrix}.$$

Hence, the direction-finding QP is given as follows:

QP(x^2, u^2) :

Minimize $\quad -3.1009624d_1 - 3.844842d_2 + \dfrac{1}{2}\left[4d_1^2 + 4d_2^2 - 4d_1d_2 \right]$

120

subject to $\quad 2.6354888d_1 - d_2 \leq 0$

$$d_1 + 5d_2 \leq 0$$
$$-0.6588722 - d_1 \leq 0$$
$$-0.8682256 - d_2 \leq 0.$$

Observe that $(d_1, d_2) = (0, 0)$ is a KKT solution for this problem with the vector of Lagrange multipliers associated with the respective constraints given by $u^3 \equiv [0.822431039, 0.933454607, 0, 0]^t$. Hence, we stop with x^2 as an optimal solution (with Lagrange multipliers u^3).

10.33 Let $F(d) \equiv \|-\nabla f(x) - d\|^2 = d^t d + 2\nabla f(x)^t d + \nabla f(x)^t \nabla f(x)$. Let $\bar{d}$ denote an optimal solution to the problem:

D: $\min \{\frac{1}{2} F(d) : A_1 d = 0\}$.

a. The function $F(d)$ is strictly convex, while the system of constraints $A_1 d = 0$ is linear and consistent. Therefore, the KKT conditions are both necessary and sufficient for optimality. A vector $\bar{d}$ is a KKT point for Problem D if $\bar{d} = -\nabla f(x) + A_1^t u$ and $A_1 \bar{d} = 0$. If $\bar{d}$ solves Problem D, then it must be a KKT point, that is:

1. It is in the nullspace, $N(A_1)$, of A_1 (since $A_1 \bar{d} = 0$).

2. It is a sum of $A_1^t u$, which is in the orthogonal complement of the nullspace of A_1, and $-\nabla f(x)$.

From linear algebra, we can therefore claim that $\bar{d}$ is the projection vector of $-\nabla f(x)$ onto the nullspace of A_1. If so, out of all vectors in $N(A_1)$, $\bar{d}$ is closest to $-\nabla f(x)$.

If $\bar{d}$ is the projection vector of $-\nabla f(x)$ onto $N(A_1)$, then

1. $\bar{d}$ must be a vector in $N(A_1)$, that is $A_1 \bar{d} = 0$.

2. There exists a unique vector $-z$ in the orthogonal complement of $N(A_1)$ such that $\bar{d} - z = -\nabla f(x)$.

121

That is, there exists a vector u such that $A_1^t u = z$, which further yields the existence of a solution to the system $A_1 \bar{d} = 0$, $-\nabla f(x) = \bar{d} - A_1^t u$. Thus, $\bar{d}$ is a KKT point.   □

b. By premultiplying the equation $\bar{d} = -\nabla f(x) + A_1^t u$ by A_1, and noting that $A_1 \bar{d} = 0$, we obtain $-A_1 \nabla f(x) + A_1 A_1^t u = 0$. If A_1 is of full row-rank, then $A_1 A_1^t$ is positive definite and hence nonsingular, and we get $u = (A_1 A_1^t)^{-1} A_1 \nabla f(x)$, so that
$$\bar{d} = [-I + A_1^t (A_1 A_1^t)^{-1} A_1] \nabla f(x).$$
(Observe that by the derivation in Part (a), we directly have $\bar{d} = -P\nabla f(x)$, where $P = I - A_1^t (A_1 A_1^t)^{-1} A_1$ is the projection matrix onto the nullspace of A_1.)   □

c. Given that $\nabla f(x) = [2 \; -3 \; 3]^t$, $A_1 = \begin{bmatrix} 2 & 2 & -3 \\ 2 & 1 & 2 \end{bmatrix}$, we get the KKT system

$$d_1 = -2 + 2u_1 + 2u_2$$
$$d_2 = 3 + 2u_1 + u_2$$
$$d_3 = -3 - 3u_1 + 2u_2$$
$$2d_1 + 2d_2 - 3d_3 = 0$$
$$2d_1 + d_2 + 2d_3 = 0.$$

This yields $(u_1, u_2) = (\dfrac{-11}{17}, \dfrac{7}{9})$, and $\bar{d} = (d_1, d_2, d_3)^t = (\dfrac{-266}{153}, \dfrac{380}{153}, \dfrac{76}{153})^t$.

10.36 a. Let $M^t = [A^t \; -E^t]$, where E^t is an $n \times (n - m)$ submatrix of I_n (the columns of I_n correspond to the nonbasic variables). By assumption, A is an $m \times n$ matrix whose rank is m. Furthermore, rank $(E^t) = n - m$, and so by construction, M^t is an $n \times n$ matrix of full rank. This implies that the matrix $M^t M$ is invertible. The projection matrix P that projects onto the nullspace of the gradients of the binding constraints is then given as $P = I - M^t(MM^t)^{-1}M$, and the direction vector d is then given as $d = -Pc$. Notice that $Md = -MPc$, which by the formula for P gives $Md = -Mc + MM^t(MM^t)^{-1}Mc = 0$. Hence, $Md = 0$, and so $d = 0$ since M is nonsingular. $\square$

b. Let $u^t = [u_0^t \ u_1^t]$, where u_0 and u_1 are $m \times 1$ and $(n-m) \times 1$ vectors, respectively. The vector u_0 is associated with the equality constraints $Ax = b$, while the vector u_1 is associated with the $n - m$ nonnegativity constraints $x_j \geq 0$ corresponding to the nonbasic variables. The equation $u = -(MM^t)^{-1}Mc$ can be rewritten as $MM^t u = -Mc$, which by the structure of the matrix M and the vector u gives the following system:

$$AA^t u_0 - AE^t u_1 = -Ac \tag{1a}$$

$$-EA^t u_0 + EE^t u_1 = Ec. \tag{1b}$$

If $A = [B \ N]$, where B is an $m \times m$ invertible matrix, and $c^t = [c_B^t \ c_N^t]$, then $AA^t = BB^t + NN^t$, $AE^t = N$, $EE^t = I_{n-m}$, and $Ec = c_N$.

Therefore, we can rewrite the system of Equation (1) as

$$BB^t u_0 + NN^t u_0 - Nu_1 = -Bc_B - Nc_N \tag{2a}$$

$$-N^t u_0 + u_1 = c_N. \tag{2b}$$

By premultiplying (2b) by the matrix N, and adding the resulting equation to (29), we obtain $BB^t u_0 = -Bc_B$. That is, $u_0 = -(B^t)^{-1}c_B$, which further gives $u_1 = c_N - N^t(B^{-1})^t c_B$. Thus, $u_0^t = -c_B^t B^{-1}$ and $u_1^t = c_N^t - c_B^t B^{-1}N$. This means that the jth entry of the vector u_1 is simply the value of the reduced cost

123

"$c_j - z_j$" used in the simplex method. Thus, the most negative u_j associated with the constraint $x_j \geq 0$ for any nonbasic variable is the most negative value of $c_j - z_j$ in the simplex method.

Given that $d' = P'c$, we have $x' = x + \lambda d'$, where $\lambda \geq 0$ denotes the optimal step length, so that we now need to show that

$$d' = \begin{bmatrix} -B^{-1}a_j \\ e_j \end{bmatrix},$$ where j is the index of entering variable in the

simplex method. By construction, $P' = I - M''(M'M'')^{-1}M'$, where $M' = \begin{bmatrix} A \\ -E' \end{bmatrix}$, and where E' is formed from E by deleting the

row corresponding to e_j^t. Let us rewrite the matrix A using three blocks: the matrix B, the column a_j, and the matrix N' (notice that $N = [a_j \ N']$). This allows us to rewrite the matrix M' as follows:

$$M' = \begin{bmatrix} B & a_j & N' \\ 0 & 0 & -1 \end{bmatrix}, \text{ where } I \text{ is of order } n - m - 1.$$

Furtheremore, let $d'' = [d_B^t \ d_j \ d_{N'}^t]$. Since $M'd' = M'P'c = 0$, we obtain

$$Bd_B + a_j d_j + N'd_{N'} = 0 \text{ and } d_{N'} = 0.$$

Taking $d_j = 1$ by way of increasing the jth nonbasic variable by a

unit, this yields $d_B = -B^{-1}a_j$ and $d_{N'} = 0$. Thus $d' = \begin{bmatrix} -B^{-1}a_j \\ e_j \end{bmatrix}$.

□

c. The above steps now match with the application of the simplex method.

10.41 a. The figure below illustrates the trajectory of iterates. The details pertaining to these iterates are as follows:

Iteration 1:

Start at the origin $x^1 = (0, 0)^t$ (with coordinates specified in the space of the (x_1, x_2)-variables). The nonbasic variables are $\{x_1, x_2\}$ and the basic variables are $\{x_3, x_4, x_5\}$. Noting the orientation of $-\nabla f(x^1)$, we get $r_1 < 0$ and $r_2 < 0$. Assuming $|r_2| > |r_1|$, we increase x_2 and perform a line search to get the new iterate x^2 as shown in the figure.

Iteration 2:

At x^2 we get the following:

Nonbasic variables: $\{x_1, x_2\}$;

Basic variables: $\{x_3, x_4, x_5\}$;

$r_1 < 0$ and $r_2 < 0$.

Hence, we increase x_1. Performing a line search produces x^3 as shown in the figure.

Subsequent iterations:

The trajectory zigzags as shown, finally obtaining an iterate x^k as shown in the figure, where we have the following:

Nonbasic variables: $\{x_1, x_5\}$;

Basic variables: $\{x_2, x_3, x_4\}$;

$r_1 > 0$ and $r_5 > 0$.

Since $x_5 = 0$ and $x_1 > 0$, we decrease x_1 to reach the optimum x^* upon performing a line search.

At x^*, the basis remains the same, but we now have $r_1 = 0$ (with $x_1 > 0$) and $r_5 > 0$ (with $x_5 = 0$), and so the solution x^* is a KKT point (and an optimum in this case).

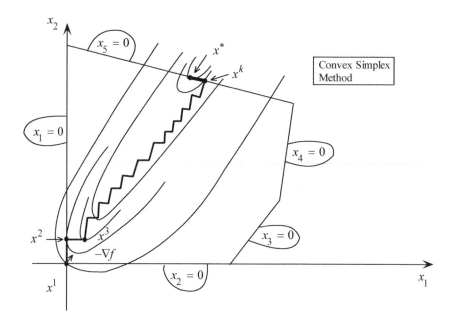

b.   The reduced gradient method would essentially imitate the steepest descent method in the initial iterations for this example by following the negative (reduced) gradient as illustrated in the figure below, until the trajectory intercepts the constraint $x_5 = 0$ as shown in the figure at the depicted point x^4. At this solution, we now have the following:

Nonbasic variables: $\{x_1, x_5\}$;

Basic variables: $\{x_2, x_3, x_4\}$;

$r_1 > 0$ and $r_5 > 0$.

Since $x_5 = 0$, the method would decrease x_1 (holding $x_5 = 0$), and thus reach the solution x^* and verify this to be optimal as in Part (a) above.

126

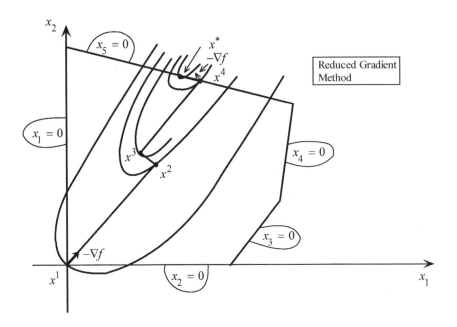

c. In this case, using the quadratic programming subproblem (10.62) in the space of the superbasic variables at the starting solution x^1, where we have $\{x_1,\ x_2\}$ as superbasic variables and $\{x_3,\ x_4,\ x_5\}$ as basic variables, the direction generated is essentially oriented toward the unconstrained optimum (of the given *quadratic* objective function). Performing a line search along this direction produces the new iterate x^2 as shown in the figure.

Now, at x^2, since the basic variable x_5 has become zero, we exchange it in the basis with x_2, say, so that we now have:

Nonbasic variables: $\{x_5\}$;

Superbasic variables: $\{x_1\}$;

Basic variables: $\{x_2,\ x_3,\ x_4\}$;

$r_1 < 0$.

The quadratic approximation in the superbasic variable space over the one-dimensional subspace defined by $x_5 = 0$ essentially generates the same direction as would be obtained by the convex simplex or the reduced gradient method in this case, and leads to the solution x^* in the next iterations, which is then verified to be a KKT point (optimal in this case) as in Parts (a) and (b) above.

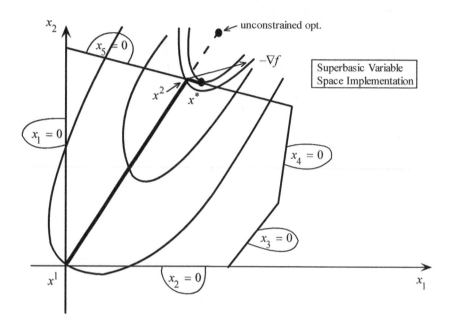

10.44 We first add slack variables x_3 and x_4 to the first two inequalities.

Start with $x^1 = [0\ 0\ 0\ 6\ 4]^t$. $I_1 = \{3,\ 4\}$

Iteration 1:

	nb	nb	b	b
	x_1	x_2	x_3	x_4
Solution x^1	0	0	6	4
$\nabla f(x^1)$	3	−4	0	0
$\nabla_B f(x^1) = \begin{bmatrix} 0 \\ 0 \end{bmatrix}$	3	2	1	0
	−1	2	0	1
r	3	−4	0	0

$\alpha = 4,\ \beta = 0;\ \alpha > \beta;\ v = 2;\ d = [0\ 1\ -2\ -2]^t;$

$\lambda_{max} = \min\{6/2,\ 4/2\} = 2;\ \lambda_1 = \min\{2,\ \sqrt{\frac{2}{3}}\} = \sqrt{\frac{2}{3}}$. Hence,

$$x^2 = \left[0,\ \frac{\sqrt{2}}{\sqrt{3}},\ 6 - \frac{2\sqrt{2}}{\sqrt{3}},\ 4 - \frac{2\sqrt{2}}{\sqrt{3}}\right]^t.$$

Iteration 2: $I_2 = \{3,\ 4\}$.

	nb x_1	nb x_2	b x_3	b x_4
Solution x^2	0	$\sqrt{2}/\sqrt{3}$	$6 - \dfrac{2\sqrt{2}}{\sqrt{3}}$	$4 - \dfrac{2\sqrt{2}}{\sqrt{3}}$
$\nabla f(x^2)$	3	0	0	0
	1	2	1	0
$\nabla_B f(x^2) = \begin{bmatrix} 0 \\ 0 \end{bmatrix}$	−1	2	0	1
r	3	0	0	0

We now get $\alpha = 0$ and $\beta = 0$, and so we stop. The solution $\bar{x}_1 = 0$, $\bar{x}_2 = \sqrt{2}/\sqrt{3}$ is a KKT point. Note that the Hessian $H(x) = \begin{bmatrix} 6x_1 - 2 & 0 \\ 0 & 12x_2 \end{bmatrix}$ is indefinite over the feasible region, and so is $H(\bar{x})$. However, the KKT system for Problem P′ to minimize $f(x)$ over the relaxed nonnegative region $x \geq 0$ is given by

$$3x_1^2 - 2x_1 + 3 - u_1 = 0 \tag{1}$$

$$6x_2^2 - 4 - u_2 = 0 \tag{2}$$

$$(x_1,\ x_2) \geq 0,\ (u_1,\ u_2) \geq 0,\ x_1 u_1 = x_2 u_2 = 0. \tag{3}$$

Multiplying (1) by x_1 and (2) by x_2 and using (3) yields

$$3x_1^3 - 2x_1^2 + 3x_1 = 0 \tag{4}$$

$$6x_2^3 - 4x_2 = 0. \tag{5}$$

Hence, (4) yields $x_1 = 0$ as the only real root, which from (1) yields the Lagrange multiplier $u_1 = 3$. Also, from (5), we get that either $x_2 = 0$, which means that $u_2 = -4$ from (2), or $x_2 = \sqrt{2}/\sqrt{3}$, which yields $u_2 = 0$. Hence, the *unique* KKT solution to Problem P' is $\bar{x} = (0, \sqrt{2}/\sqrt{3})$ with Lagrange multipliers (3, 0), and since the KKT conditions are *necessary* for Problem P' (because the constraints are linear), we get that $\bar{x}$ uniquely solves Problem P'. But $\bar{x}$ is also a feasible (KKT) solution to the given problem, which is a restriction of Problem P', and so $\bar{x}$ is the unique global optimum for the given problem.

10.47 a. By the given formula, we have that $d = 0 \Leftrightarrow r_j \geq 0$ for all $j \in J$, with $r_j = 0$ if $x_j > 0$, where J is the index set of the nonbasic variables. By Theorem 10.6.1, this happens if and only if the given point x is a KKT point. (Note that this condition implies that $\min\{r_N^t d_N : d_j \geq 0$ if $x_j = 0$ for $j \in J\} = 0$.)

b. By Part (a), if $\bar{d} \neq 0$ at some $\bar{x}$, then $\bar{d}_N \neq 0$ and we have $\nabla f(\bar{x})^t \bar{d} = r_N^t \bar{d}_N < 0$ where $d_N = \bar{d}_N$ solves the following problem:

$$\min\{r_N^t d_N : d_j \geq 0 \text{ if } x_j = 0 \text{ for } j \in J\}. \tag{1}$$

Hence, $\bar{d}$ is an improving direction. Furthermore, since $\bar{x}_B > 0$ and since $\bar{d}_j \geq 0$ if $\bar{x}_j = 0$ for $j \in J$, we have that $\bar{d}$ is also a feasible direction (with $\bar{d}_B = -B^{-1} N \bar{d}_N$).

c. Let $f(x) \equiv 3e^{-2x_1 + x_2} + 2x_1^2 + 2x_1 x_2 + 3x_2^2 + x_1 + 3x_2$.

Denote x_3 and x_4 as the slack variables in the two structural inequality constraints, respectively, and let x^k and d^k be the iterates and the search direction, respectively, at iteration k. Thus,

$$\nabla f(x) = \begin{bmatrix} -6e^{-2x_1 + x_2} + 4x_1 + 2x_2 + 1 \\ 3e^{-2x_1 + x_2} + 2x_1 + 6x_2 + 3 \\ 0 \\ 0 \end{bmatrix}.$$

Iteration 1.

$x^1 = (0, 0, 4, 3)^t$ with $f(x^1) = 3$.

$\nabla f(x^1) = (-5, 6, 0, 0)^t$.

Nonbasic variables: $(x_1, x_2) = (0, 0)$

Basic variables: $(x_3, x_4) = (4, 3)$.

	nb x_1	nb x_2	b x_3	b x_4
Solution x^1	0	0	4	3
$\nabla f(x^1)$	-5	6	0	0
$\nabla_B f(x^1) = \begin{bmatrix} 0 \\ 0 \end{bmatrix}$	2	1	1	0
	-1	1	0	1
r	-5	6	0	0

Hence, $d_N = \begin{bmatrix} 5 \\ 0 \end{bmatrix}$ and $d_B = -5\begin{bmatrix} 2 \\ -1 \end{bmatrix}$. Thus, $d^1 = [5, 0, -10, 5]^t$,

with $\lambda_{max} = 4/10 = 0.4$. The line search problem is given as follows:

minimize $\{f(x^1 + \lambda d^1) : 0 \leq \lambda \leq \lambda_{max}\}$,

which yields $\lambda^* = 0.08206$. This leads to the new iterate

$x^2 = (0.4103, 0, 3.1794, 3.4103)^t$ with $f(x^2) = 2.0675$.

Retaining the same basis, we get the following:

Iteration 2.

	nb x_1	nb x_2	b x_3	b x_4
Solution x^2	0.4103	0	3.1794	3.4103
$\nabla f(x^2)$	0	2.5	0	0
$\nabla_B f(x^1) = \begin{bmatrix} 0 \\ 0 \end{bmatrix}$	2	1	1	0
	-1	1	0	1
r	0	2.5	0	0

Since $r_1 = 0$ and $r_2 > 0$ with $x_2 = 0$, the current solution x^2 is a KKT point. Because f is strictly convex (the Hessian of f is PD in the (x_1, x_2)-space), we get that x^2 is the unique global minimum.

d. Consider the following problem:

Minimize $\quad -x_1 - x_2$

subject to $\quad x_1 + x_3 = 1$

$\qquad\qquad x_2 + x_4 = 1$

$\qquad\qquad (x_1, x_2, x_3, x_4) \geq 0.$

Let $x^k = (0, 1 - \dfrac{1}{k}, 1, \dfrac{1}{k})^t$ for $k \geq 3$, so that we have $\{x_2, x_3\}$ as basic variables, and $\{x_1, x_4\}$ as nonbasic variables. At Iteration k, we have the following:

			nb x_1	b x_2	b x_3	nb x_4
Solution x^k			0	$1 - \dfrac{1}{k}$	1	$\dfrac{1}{k}$
$\nabla f(x^k)$			-1	-1	0	0
$\nabla_B f(x^k) = \begin{bmatrix} 0 \\ -1 \end{bmatrix}$	**Basic:**	x_3	1	0	1	0
		x_2	0	1	0	1
r			-1	0	0	1

Thus, $d_N = -r_N = \overset{x_1 \quad x_4}{[1 \quad -1]^t}$, and $d_B = 1\begin{bmatrix} -1 \\ 0 \end{bmatrix} - 1\begin{bmatrix} 0 \\ -1 \end{bmatrix}$, i.e.,

$d_B = \overset{x_3 \quad x_2}{[-1 \quad 1]^t}$

Hence, $d^k = [1, 1, -1, -1]^t$.

As $k \to \infty$, we get $x^k \to \bar{x} \equiv [0, 1, 1, 0]^t$ and

132

$d^k \rightarrow \bar{d} = [1, 1, -1, -1]^t$. However, in the limit at $\bar{x}$, we have that $\bar{d}$ is not a feasible direction since $\bar{x}_4 = 0$ and $\bar{d}_4 = -1$. In fact, for $\bar{x}$, the algorithm would use the same basis as above and compute the same **r**-vector, but since $\bar{x}_4 = 0$ and $r_4 > 0$, it would select

$$d_N = \overset{x_1 \ x_4}{[1 \ \ 0]^t}, \text{ and so } d_B = \overset{x_3 \ x_2}{[-1 \ \ 0]^t}, \text{ thus yielding the direction}$$

$d = [1, 0, -1, 0]^t \neq \bar{d}$. Hence, the given direction-finding map is not closed.

10.52 Since $Bx_B + Nx_N = b$, we obtain $x_B = B^{-1}b - B^{-1}Nx_N$. To guarantee that all variables take on nonnegative values only, we need to require that $B^{-1}b - B^{-1}Nx_N \geq 0$ and $x_N \geq 0$. Furthermore, by substituting $B^{-1}b - B^{-1}Nx_N$ for x_B into the objective function we obtain an equivalent problem $P(x_N)$ in the nonbasic variable space.

a. At the current solution $\bar{x}_N$, letting $J_{nb} \equiv \{j : x_j \text{ is nonbasic}\}$, $J_0 \equiv \{j \in J_{nb} : \bar{x}_j = 0\}$, and $J_+ \equiv \{j \in J_{nb} : \bar{x}_j > 0\}$, we have (noting that $\bar{x}_B > 0$), the set of binding constraints are given by $x_j \geq 0$ for $j \in J_0$. Furthermore, by the chain rule for differentiation, we obtain

$$\nabla F(x_N)^t = -\nabla_B f(B^{-1}b - B^{-1}Nx_N, \ x_N)^t B^{-1}N$$
$$+\nabla_N f(B^{-1}b - B^{-1}Nx_N, \ x_N)^t.$$

From the above equation it follows that $\nabla F(x_N) = r_N$ in the reduced gradient method.

b. The KKT system for $P(x_N)$ is given as follows:

$$-(B^{-1}N)^t \nabla_B f(B^{-1}b - B^{-1}Nx_N, \ x_N)$$
$$+\nabla_N f(B^{-1}b - B^{-1}Nx_N, \ x_N) + (B^{-1}N)^t u - w = 0;$$
$$B^{-1}b - B^{-1}Nx_N \geq 0;$$
$$x_N \geq 0, \ u \geq 0, \ w \geq 0;$$

133

$$u^t(B^{-1}b - B^{-1}Nx_N) = 0, \quad x_N^t w = 0.$$

Let $\bar{u}$ and $\bar{w}$ denote the Lagrange multipliers for the above system at $x_N = \bar{x}_N$. Since $\bar{x}_B = B^{-1}b - B^{-1}N\bar{x}_N > 0$, we necessarily have $\bar{u} = 0$. Therefore, at $\bar{x}_N$, we have

$$\bar{w} = \nabla_N f(B^{-1}b - B^{-1}N\bar{x}_N,\ \bar{x}_N)$$
$$-(B^{-1}N)^t \nabla_B f(B^{-1}b - B^{-1}N\bar{x}_N,\ \bar{x}_N),$$

so that the KKT system reduces to the following:

$$\nabla_N f(B^{-1}b - B^{-1}N\bar{x}_N,\ \bar{x}_N) - (B^{-1}N)^t \nabla_B f(B^{-1}b - B^{-1}N\bar{x}_N,\ \bar{x}_N) \geq 0;$$
$$\bar{x}_N^t[\nabla_N f(B^{-1}b - B^{-1}N\bar{x}_N,\ \bar{x}_N) - (B^{-1}N)^t \nabla_B f(B^{-1}b - B^{-1}N\bar{x}_N,\ \bar{x}_N)] = 0,$$

or more concisely,

$$r_N \geq 0 \text{ and } \bar{x}_N^t r_N = 0,$$
where

$$r_N = \nabla_N f(B^{-1}b - B^{-1}N\bar{x}_N,\ \bar{x}_N) - (B^{-1}N)^t \nabla_B f(B^{-1}b - B^{-1}N\bar{x}_N,\ \bar{x}_N) \geq 0.$$

The foregoing necessary and sufficient conditions for $\bar{x}_N$ to be a KKT point for $P(x_N)$ are equivalent to $r_j \geq 0$ for $j \in J_0$, and $r_j = 0$ for $j \in J_+$, which in turn are identical to the stopping rule used in the reduced gradient method (Theorem 10.6.1). Hence, this gives necessary and sufficient conditions for $(\bar{x}_B,\ \bar{x}_N)$ to be a KKT point for the original problem.

CHAPTER 11:

LINEAR COMPLEMENTARY PROBLEM, AND QUADRATIC, SEPARABLE, FRACTIONAL, AND GEOMETRIC PROGRAMMING

11.1   a.   The KKT system for the given LP is as follows:

$$-u + A^t v = -c$$
$$Ax = b$$
$$u^t x = 0, \ x \geq 0, \ u \geq 0.$$

b.   The KKT system for the problem given in Part (b) is as follows:

$$2u_1 - u_2 - u_3 = 1$$
$$3u_1 + 2u_2 \qquad -u_4 = 3$$
$$2x_1 + 3x_2 + y_1 \qquad = 6$$
$$-x_1 + 2x_2 \qquad +y_2 = 2,$$
$$x_1 u_3 = 0, \ x_2 u_4 = 0, \ u_1 y_1 = 0, \ u_2 y_2 = 0$$
$$x_1 \geq 0, \ x_2 \geq 0, \ y_1 \geq 0, \ y_2 \geq 0, \ u_i \geq 0 \text{ for } i = 1, 2, 3, 4.$$

Let $w = [u_3 \ u_4 \ y_1 \ y_2]^t$, $z = [x_1 \ x_2 \ u_1 \ u_2]^t$, $q = [-1 \ -3 \ 6 \ 2]^t$,

and $M = \begin{bmatrix} 0 & 0 & 2 & -1 \\ 0 & 0 & 3 & 2 \\ -2 & -3 & 0 & 0 \\ 1 & -2 & 0 & 0 \end{bmatrix}$.

Then the KKT system can be rewritten as the following linear complementarity problem: $w - Mz = q, \ w \geq 0, \ z \geq 0, \ w^t z = 0$.

The complementary pivoting algorithm then proceeds as follows:

	w_1	w_2	w_3	w_4	z_1	z_2	z_3	z_4	z_0	RHS
w_1	1	0	0	0	0	0	−2	1	−1	−1
w_2	0	1	0	0	0	0	−3	−2	⊘−1	−3
w_3	0	0	1	0	2	3	0	0	−1	6
w_4	0	0	0	1	−1	2	0	0	−1	2

	w_1	w_2	w_3	w_4	z_1	z_2	z_3	z_4	z_0	RHS
w_1	1	−1	0	0	0	0	1	3	0	2
z_0	0	−1	0	0	0	0	3	2	1	3
w_3	0	−1	1	0	2	3	3	2	0	9
w_4	0	−1	0	1	−1	②	3	2	0	5
w_1	1	−1	0	0	0	0	1	③	0	2
z_0	0	−1	0	0	0	0	3	2	1	3
w_3	0	1/2	1	−3/2	7/2	0	−3/2	−1	0	3/2
z_2	0	−1/2	0	1/2	−1/2	1	3/2	1	0	5/2
z_4	1/3	−1/3	0	0	0	0	1/3	1	0	2/3
z_0	−2/3	−1/3	0	0	0	0	7/3	0	1	5/3
w_3	1/3	1/6	1	−3/2	⑦/2	0	−7/6	0	0	13/6
z_2	−1/3	−1/6	0	1/2	−1/2	1	7/6	0	0	11/6
z_4	1/3	−1/3	0	0	0	0	1/3	1	0	2/3
z_0	−2/3	−1/3	0	0	0	0	⑦/3	0	1	5/3
z_1	2/21	1/21	2/7	−3/7	1	0	−1/3	0	0	13/21
z_2	−2/7	−1/7	1/7	2/7	0	1	1	0	0	15/7
$u_2 \equiv z_4$	3/7	−2/7	0	0	0	0	0	1	−1/7	3/7
$u_1 \equiv z_3$	−2/7	−1/7	0	0	0	0	1	0	3/7	5/7
$x_1 \equiv z_1$	0	0	2/7	−3/7	1	0	0	0	1/7	6/7
$x_2 \equiv z_2$	0	0	1/7	2/7	0	1	0	0	−3/7	10/7

Because $z_0 = 0$, the algorithm stops with a solution to the linear complementarity problem obtained as follows: $x_1 = 6/7$, $x_2 = 10/7$, $u_1 = 5/7$, $u_2 = 3/7$, and all the remaining variables equal to zero.

c. For the modified LP, the KKT system is given as follows:

$$u_2 + u_3 = -1$$
$$-u_1 - 2u_2 + u_4 = -3$$
$$x_2 + y_1 = 2$$
$$-x_1 + 2x_2 + y_2 = 2$$
$$(x_1,\ x_2,\ y_1,\ y_2,\ u_1,\ u_2,\ u_3,\ u_4) \geq 0$$

136

$$x_1 u_3 = x_2 u_4 = y_1 u_1 = y_2 u_2 = 0.$$

Note that the first constraint implies (with $u \geq 0$) that the dual is infeasible, and since the primal is feasible, we have that the primal is unbounded. Hence, no complementarity solution exists and we expect Lemke's algorithm to stop with a ray termination. Putting the above KKT system in a standard tableau (while keeping the columns of complementary pairs of variables adjacent for convenience, we obtain the following tableaus along with the shown sequence of pivots:

	x_1	u_3	x_2	u_4	y_1	u_1	y_2	u_2	z_0	RHS
u_3	0	1	0	0	0	0	0	1	−1	−1
u_4	0	0	0	1	0	−1	0	−2	(−1)	−3
y_1	0	0	1	0	1	0	0	0	−1	2
y_2	−1	0	2	0	0	0	1	0	−1	2
u_3	0	1	0	−1	0	1	0	3	0	2
z_0	0	0	0	−1	0	1	0	2	1	3
y_1	0	0	1	−1	1	1	0	2	0	5
y_2	−1	0	(2)	−1	0	1	1	2	0	5
u_3	0	1	0	−1	0	1	0	(3)	0	2
z_0	0	0	0	−1	0	1	0	2	1	3
y_1	1/2	0	0	−1/2	1	1/2	−1/2	1	0	5/2
x_2	−1/2	0	1	−1/2	0	1/2	1/2	1	0	5/2
u_2	0	1/3	0	−1/3	0	1/3	0	1	0	2/3
z_0	0	−2/3	0	−1/3	0	1/3	0	0	1	5/3
y_1	(1/2)	−1/3	0	−1/6	1	1/6	−1/2	0	0	11/6
x_2	−1/2	−1/3	1	−1/6	0	1/6	1/2	0	0	11/6
u_2	0	1/3	0	−1/3	0	(1/3)	0	1	0	2/3
z_0	0	−2/3	0	−1/3	0	1/3	0	0	1	5/3
x_1	1	−2/3	0	−1/3	2	1/3	−1	0	0	11/3
x_2	0	−2/3	1	−1/3	1	1/3	0	0	0	11/3
	x_1	u_3	x_2	u_4	y_1	u_1	y_2	u_2	z_0	RHS

u_1	0	1	0	-1	0	1	0	3	0	2
z_0	0	-1	0	0	0	0	0	-1	1	1
x_1	1	-1	0	0	2	0	-1	-1	0	3
x_2	0	-1	1	0	1	0	0	-1	0	3

At this point, since u_2 just left the basis at the previous iteration, we enter y_2 and achieve ray termination. By Theorem 11.2.4 (with $H \equiv 0$), since the primal is feasible (verified by Phase I or graphically in this case), we conclude that the primal is unbounded.

11.5 Let P and PKKT denote the following given problems:

P: Minimize $f(x) = c^t x + \frac{1}{2} x^t H x$

subject to $Ax = b$

$x \geq 0,$

and

PKKT: Minimize $g(x, u, v) = c^t x - b^t u$

subject to $Ax = b$

$Hx + A^t u - v = -c$

$v^t x = 0$

$x \geq 0, \ v \geq 0, \ u$ unrestricted.

a. From the KKT system formulation for Problem P we can assert that (x, u, v) is a feasible solution for Problem PKKT if and only if x is a KKT point for Problem P. Furthermore, if a triple (x, u, v) is feasible for PKKT, then we necessarily have $u^t Ax = u^t b$, and $x^t Hx + x^t A^t u = -x^t c$, which yields $x^t Hx = -(c^t x + b^t u)$. Therefore, for any such (x, u, v), the objective function $f(x)$ can be rewritten as

$$f(x) = c^t x - \frac{1}{2}(c^t x + b^t u) = \frac{1}{2}(c^t x - b^t u) = \frac{1}{2} g(x, u, v).$$

Therefore, an optimal solution for Problem PKKT is a KKT point for Problem P having the smallest value of the function $f(x)$.

Furthermore, since the KKT conditions are necessary for Problem P, we can pose the global optimization of Problem P as finding a KKT solution having the least objective value. But this is precisely what

138

Problem PKKT accomplishes. Hence, for any symmetric matrix H, Problem PKKT determines a global optimum for Problem P.

b. Evident from Part (a). Also, based on Dorn's duality (see Chapter 6), this is like minimizing the difference between the primal and dual objective values over the primal and dual space, in addition to complementary slackness, based on the dual problem written as

$$\min\{\frac{1}{2}x^t Hx + u^t b : Hx + A^t u - v = -c, \ v \geq 0, \ u \text{ unrestricted}\}.$$

c. For the given problem, we have the following:

P: Minimize $f(x) =$

$$-(x_1 - 2)^2 - (x_2 - 2)^2 = -x_1^2 - x_2^2 + 4x_1 + 4x_2 - 8$$

subject to $-2x_1 + x_2 + x_3 = 4$

$$3x_1 + 2x_2 + x_4 = 12$$

$$3x_1 - 2x_2 + x_5 = 6$$

$$(x_1, \ x_2, \ x_3, \ x_4, \ x_5) \geq 0$$

and (with the objective written as $\frac{1}{2}g \equiv f$), we have,

PKKT: Minimize $\frac{1}{2}[4x_1 + 4x_2 - 4u_1 - 12u_2 - 6u_3] - 8$

subject to $-2x_1 + x_2 + x_3 = 4$

$$3x_1 + 2x_2 + x_4 = 12$$

$$3x_1 - 2x_2 + x_5 = 6$$

$$-2x_1 - 2u_1 + 3u_2 + 3u_3 - v_1 = -4$$

$$-2x_2 + u_1 + 2u_2 - 2u_3 - v_2 = -4$$

$$x_1 v_1 = x_2 v_2 = x_3 u_1 = x_4 u_2 = x_5 u_3 = 0 \qquad (1)$$

$$(x, u, v) \geq 0.$$

As per Exercise 11.4, we can solve PKKT using a branch-and-bound algorithm based on the dichotomy that one or the other of each complementary pair of variables is zero. However, in this case, viewing the problem in the $(x_1, \ x_2)$-space and noting that the objective function of Problem P is concave, we must have an extreme point optimum. Graphically, the optimum to Problem P is given by

$(x_1^*, x_2^*) = (\frac{4}{7}, \frac{36}{7})$, with $(x_3^*, x_4^*, x_5^*) = (0, 0, \frac{102}{7})$, $(v_1^*, v_2^*) =$

$(0, 0)$, and $(u_1^*, u_2^*, u_3^*) = (\frac{172}{49}, \frac{68}{49}, 0)$, with objective value (in P

and PKKT) being equal to $\dfrac{-584}{49}$.

11.12 a. Let 1_p denote a $p \times 1$ vector of ones. An equilibrium pair $(\bar{x}, \bar{y})$
for a bimatrix game with loss matrices A and B can be characterized
as follows:

$\bar{x}$ solves **P1:** Minimize $\quad x^t A \bar{y}$

$\qquad\qquad$ subject to $\quad -1_m^t x \geq -1 \quad \leftarrow v_1 \qquad\qquad$ (1)

$\qquad\qquad\qquad\qquad\quad 1_m^t x \geq 1 \quad \leftarrow v_2 \qquad\qquad$ (2)

$\qquad\qquad\qquad\qquad\quad x \geq 0$

$\bar{y}$ solves **P2:** Minimize $\quad x^t B y$

$\qquad\qquad$ subject to $\quad -1_n^t y \geq -1 \quad \leftarrow v_3 \qquad\qquad$ (3)

$\qquad\qquad\qquad\qquad\quad 1_n^t y \geq 1 \quad \leftarrow v_4 \qquad\qquad$ (4)

$\qquad\qquad\qquad\qquad\quad y \geq 0$

where (v_1, v_2) and (v_3, v_4) are dual variables associated with the
constraints of P1 and P2, respectively, and where the respective dual
problems are given as follows:

D1: Maximize $\quad v_2 - v_1$

$\qquad$ subject to $\quad -1_m v_1 + 1_m v_2 \leq A\bar{y} \qquad\qquad$ (5)

$\qquad\qquad\qquad\quad (v_1, v_2) \geq 0$

D2: Maximize $\quad v_4 - v_3$

$\qquad$ subject to $\quad -1_n v_3 + 1_n v_4 \leq B^t \bar{x} \qquad\qquad$ (6)

$\qquad\qquad\qquad\quad (v_3, v_4) \geq 0.$

Writing the primal-dual feasibility and complementary slackness
necessary and sufficient optimality conditions for the above pair of
primal and dual problems, and denoting surplus variables s_1, s_2,
s_3, and s_4 for the constraints (1) – (4), respectively, and slack
variables w_1 and w_2 for the constraints (5) and (6), respectively, we

have that $(\bar{x}, \bar{y})$ is an equilibrium pair if and only if there exist nonnegative variables v_1, v_2, v_3, v_4, s_1, s_2, s_3, s_4, w_1, and w_2 such that

$$-A\bar{y} - 1_m v_1 + 1_m v_2 + w_1 = 0$$

$$-B^t\bar{x} - 1_n v_3 + 1_n v_4 + w_2 = 0$$

$$1_m^t \bar{x} + s_1 = 1$$

$$-1_m^t \bar{x} + s_2 = -1$$

$$1_n^t \bar{y} + s_3 = 1$$

$$-1_n^t \bar{y} + s_4 = -1$$

$$\bar{x} \geq 0,\ \bar{y} \geq 0,\ (s_1,\ s_2,\ s_3,\ s_4) \geq 0,$$

$$(v_1,\ v_2,\ v_3,\ v_4) \geq 0,\ w_1 \geq 0,\ w_2 \geq 0,$$

$$s_1 v_1 = s_2 v_2 = s_3 v_3 = s_4 v_4 = \bar{x}^t w_1 = \bar{y}^t w_2 = 0,$$

where $\bar{x} \in R^m$, $\bar{y} \in R^n$, $w_1 \in R^m$, $w_2 \in R^n$, $\{s_1, s_2, s_3, s_4, v_1, v_2, v_3, v_4\} \subseteq R$, and where A and B are $m \times n$ matrices. The above system represents a linear complementarity problem of the type

$$w - Mz = q,\ w \geq 0,\ z \geq 0,\ \text{and}\ w^t z \geq 0, \tag{7}$$

where $w^t = [w_1^t\ w_2^t\ s_1\ s_2\ s_3\ s_4]$, $z^t = [\bar{x}^t\ \bar{y}^t\ v_1\ v_2\ v_3\ v_4]$, $q^t = [0^t\ 0^t\ 1\ -1\ 1\ -1]$, and

$$
M = \begin{bmatrix}
0 & A & 1_m & -1_m & 0 & 0 \\
B^t & 0 & 0 & 0 & 1_n & -1_n \\
-1_m^t & 0 & 0 & 0 & 0 & 0 \\
1_m^t & 0 & 0 & 0 & 0 & 0 \\
0 & -1_n^t & 0 & 0 & 0 & 0 \\
0 & 1_n^t & 0 & 0 & 0 & 0
\end{bmatrix}
$$

and where 0 denotes a zero matrix of appropriate order.

141

b. For any $z^t = [\bar{x}^t \ \bar{y}^t \ v_1 \ v_2 \ v_3 \ v_4]$ we have $z^t M z = \bar{x}^t A \bar{y} + 1^t_m \bar{x}(v_1 - v_2) + \bar{x}^t B \bar{y} + 1^t_n \bar{y}(v_3 - v_4) + 1^t_m \bar{x}(v_2 - v_1) + 1^t_n \bar{y}(v_4 - v_3) = \bar{x}^t (A + B)\bar{y}$. Hence, the matrix M is copositive if the matrix $A + B$ is nonnegative. However, even then, the matrix M is not necessarily copositive plus. To see this, note that $(M + M^t)z = [\bar{y}^t (A^t + B^t) \ \bar{x}^t (A + B) \ 0 \ 0 \ 0 \ 0]$. Hence,

$(M + M^t)z \neq 0$ if and only if $[\bar{y}^t (A^t + B^t), \ \bar{x}^t (A + B)] \neq 0$. However, for example, consider the following matrix $A + B$:

$$A + B = \begin{bmatrix} 2 & 4 \\ 0 & 10 \end{bmatrix}.$$ Here for $\bar{x}^t = [0 \ 1]$ and $\bar{y}^t = [1 \ 0]$, we have

$z^t M z = \bar{x}^t (A + B)\bar{y} = 0$, with $z \geq 0$ (for any (v_1, v_2, v_3, v_4) ≥ 0), but $\bar{x}^t (A + B) = [0 \ 10]$ and $\bar{y}^t (A^t + B^t) = [2 \ 0]$, i.e., $[M + M^t]z \neq 0$, which means that the matrix M in this example is not copositive plus. Nonetheless, for every bimatrix game, there exists an equilibrium pair as proven in *Game Theory* by Guillermo Owen (Academic Press, Second Edition, 1982).

c. For the given matrices A and B, the equilibrium-finding pairs of LPs P1 and P2 from Part (a) are given as follows:

$(\bar{x}_1, \bar{x}_2)$ solves **P1:**

Minimize $x_1(3\bar{y}_1 + 2\bar{y}_2 + 3\bar{y}_3) + x_2(\bar{y}_1 + 3\bar{y}_2 + 4\bar{y}_3)$

subject to $x_1 + x_2 = 1 \leftarrow u_1$

$\quad\quad (x_1, x_2) \geq 0.$

$(\bar{y}_1, \bar{y}_2, \bar{y}_3)$ solves **P2:**

Minimize $(2\bar{x}_1 + 3\bar{x}_2)y_1 + (4\bar{x}_1 + 2\bar{x}_2)y_2 + (3\bar{x}_1 + \bar{x}_2)y_3$

subject to $y_1 + y_2 + y_3 = 1 \leftarrow u_2$

$\quad\quad (y_1, y_2, y_3) \geq 0.$

The duals D1 and D2 to Problems P1 and P2 are respectively given as follows:

D1: Maximize u_1

subject to $u_1 \leq 3\bar{y}_1 + 2\bar{y}_2 + 3\bar{y}_3 \leftarrow x_1$

$\quad\quad\quad u_1 \leq \bar{y}_1 + 3\bar{y}_2 + 4\bar{y}_3 \leftarrow x_2$

D2:   Maximize u_2

subject to $u_2 \le 2\bar{x}_1 + 3\bar{x}_2 \leftarrow y_1$

$u_2 \le 4\bar{x}_1 + 2\bar{x}_2 \leftarrow y_2$

$u_2 \le 3\bar{x}_1 + \bar{x}_2 \leftarrow y_3$

We can formulate an LCP as in Part (a) and use the complementary pivoting algorithm to determine an equilibrium pair. However, assuming that $\bar{x} > 0$, we get by complementary slackness from D1 that $u_1 = 3\bar{y}_1 + 2\bar{y}_2 + 3\bar{y}_3 = \bar{y}_1 + 3\bar{y}_2 + 4\bar{y}_3$, which together with $\bar{y}_1 + \bar{y}_2 + \bar{y}_3 = 1$ yields $\bar{y}_1 = 1/3$ and $(\bar{y}_2 + \bar{y}_3) = 2/3$. It can be verified that $\bar{y} > 0$ gives a contradiction from D2. However, putting $\bar{y}_2 = 0$, we get $(\bar{y}_1, \bar{y}_2, \bar{y}_3) = (\frac{1}{3}, 0, \frac{2}{3})$ and $u_1 = 3$, which by complementary slackness in D2 and primal feasibility in P1 yields $u_2 = 2\bar{x}_1 + 3\bar{x}_2 = 3\bar{x}_1 + \bar{x}_2$ and $\bar{x}_1 + \bar{x}_2$ $=$ 1,   i.e.,

$(\bar{x}_1, \bar{x}_2) = (\frac{2}{3}, \frac{1}{3})$, with $\bar{u}_2 = \frac{7}{3}$. Now, it can be verified that

$(\bar{x}_1, \bar{x}_2) = (\frac{2}{3}, \frac{1}{3})$, $\bar{u}_1 = 3$, $(\bar{y}_1, \bar{y}_2, \bar{y}_3) = (\frac{1}{3}, 0, \frac{2}{3})$, $\bar{u}_2 = \frac{7}{3}$

satisfy the primal-dual optimality conditions associated with the pair of problems {P1, D1} and {P2, D2}. Thus, this yields an equilibrium pair for the given bimatrix game.

11.18 a.   Suppose that the matrix $\begin{bmatrix} H & A^t \\ A & 0 \end{bmatrix}$ is singular. Then there exists a nonzero vector $[x^t \ y^t]$, where $x \in R^n$ and $y \in R^m$, such that

$$Hx + A^t y = 0 \qquad\qquad (1)$$
$$Ax \quad\;\; = 0. \qquad\qquad (2)$$

Premultiplying (1) by x^t and (2) by y^t, we obtain

$$x^t Hx + x^t A^t y = 0$$
$$y^t Ax \qquad\quad = 0, \text{ i.e., } x^t A^t y = 0,$$

which yields $x^t Hx = 0$. Now, if $x = 0$ then $y \ne 0$, but (1) gives $y = 0$ via $A^t y = 0$ since A^t has full column rank, a contradiction. Hence, x

$\ne 0$. But then, by (2), we must have $x^t Hx > 0$ since the matrix H is positive definite over the nullspace of A. This contradicts the foregoing statement that $x^t Hx = 0$. Therefore, the given matrix is nonsingular. $\square$

b. The KKT system for Problem QP is given by

$$Hx + A^t v = -c \tag{3}$$
$$Ax = b. \tag{4}$$

This is of the form

$$\begin{bmatrix} H & A^t \\ A & 0 \end{bmatrix} \begin{bmatrix} x \\ v \end{bmatrix} = \begin{bmatrix} -c \\ b \end{bmatrix},$$

and therefore yields a unique solution by Part (a).

c. Consider the KKT system (3) – (4). Since the matrix H is positive definite and thus nonsingular, then Equation (3) yields $x = -H^{-1}(c + A^t v)$. To compute v, we substitute this expression for x into Equation (4) to obtain $AH^{-1}A^t v = -(AH^{-1}c + b)$. Since H is positive definite and A has full row rank, $AH^{-1}A^t$ is also positive definite and thus nonsingular. This gives $v = -(AH^{-1}A^t)^{-1}(AH^{-1}c + b)$. Therefore,

$$x = -H^{-1}c + H^{-1}A^t(AH^{-1}A^t)^{-1}(AH^{-1}c + b).$$

Because the KKT conditions are necessary and sufficient for QP, this yields the unique optimal solution to QP.

11.19 a. Let $x = x_k + d$. Then

$$c^t x + \frac{1}{2}x^t Hx = c^t d + \frac{1}{2}d^t Hd + x_k^t Hd + c^t x_k + \frac{1}{2}x_k^t Hx_k, \tag{1}$$

and for each $i \in W_k$, we have

$$A_i^t x - b_i = A_i^t x_k + A_i^t d - b_i = A_i^t d. \tag{2}$$

Thus, from (1) and (2), the problem given in (11.74) is equivalent to Problem $QP(x_k)$, after dropping the constant term $c^t x_k + \frac{1}{2} x_k^t H x_k$ in (1) from the objective function. Therefore, if d_k solves $QP(x_k)$, then $x_k + d_k$ solves the problem given in (11.74).

b. The necessary and sufficient KKT conditions for Problem $QP(x_k)$ imply at optimality that $c + H x_k + H d_k + \sum_{i \in W_k} A_i v_i^* = 0$, where v_i^* for $i \in W_k$ are the optimal Lagrange multipliers associated with the constraints $A_i^t d = 0$ for $i \in W_k$. If $d_k = 0$, then v_i^*, $i \in W_k$, solve the system

$$c + H x_k + \sum_{i \in E} A_i v_i^* + \sum_{i \in I_k} A_i v_i^* = 0. \tag{3}$$

But the KKT system for the original problem QP at x_k is given as follows (aside from the primal feasibility of x_k):

$$c + H x_k + \sum_{i \in E} A_i v_i + \sum_{i \in I_k} A_i v_i = 0,$$

$$v_i \geq 0 \text{ for all } i \in I_k.$$

Therefore, if $v_i^* \geq 0$ for all $i \in I_k$, then (3) implies that x_k is a KKT point for QP, and thus is optimal for this problem.

c. Consider the given feasible solution x_k. If the direction vector d_k found as an optimal solution to Problem $QP(x_k)$ is not a zero vector, then a move in the direction d_k is made. If $x_k + d_k$ is feasible, then readily the next iterate $x_{k+1} = x_k + d_k$ is a feasible solution. If $x_k + d_k$ is not a feasible solution, then we compute $x_{k+1} = x_k + \alpha_k d_k$, where α_k is the maximum step length along d_k that maintains feasibility as computed via the equation given in the exercise. To verify this, note that $A_i^t x_k = b_i$ for all $i \in W_k$, and $A_i^t d_k = 0$ for all $i \in W_k$ since d_k solves Problem $QP(x_k)$.

145

Therefore, $A_i^t(x_k + \alpha d_k) = b_i$ for all $i \in W_k$ and for any $\alpha \geq 0$. This means that all the constraints in the working set W_k remain binding at $x_k + \alpha d_k$ for any $\alpha \geq 0$, and thus also at $\alpha = \alpha_k$. On the other hand, for any $i \notin I_k$, we either have $A_i^t d_k \leq 0$ or $A_i^t d_k > 0$. In the former case, this ith constraint will be satisfied for any step length $\alpha \geq 0$ since $A_i^t x_k \leq b_i$. For the latter case, feasibility requires that $A_i^t(x_k + \alpha d_k) \leq b_i$, or that

$$\alpha \leq \frac{b_i - A_i^t x_k}{A_i^t d_k} \quad \text{(since } A_i^t d_k > 0). \tag{4}$$

Thus, the maximum step length α_k to maintain feasibility of $x_{k+1} \equiv x_k + \alpha_k d_k$ is given from (4) and the foregoing argument as follows:

$$\alpha_k = \underset{i \notin I_k : A_i^t d_k > 0}{\text{minimum}} \left\{ \frac{b_i - A_i^t x_k}{A_i^t d_k} \right\}, \tag{5}$$

which is achieved at some $i = q$ as stated in the exercise. Thus we have demonstrated that if the algorithm is initialized with a feasible solution x_1, then feasibility is maintained throughout the algorithm.

Next, we show that if $d_k \neq 0$, then $f(x_{k+1}) < f(x_k)$. Notice that $d = 0$ is a feasible solution for Problem $QP(x_k)$. Therefore, by the optimality of d_k and its uniqueness (by Exercise 11.18), we have

$$\nabla f(x_k)^t d_k + \frac{1}{2} d_k^t H d_k < 0, \tag{6}$$

where $f(x)$ is the objective function for Problem QP. But $\frac{1}{2} d_k^t H d_k$ > 0 since H is PD and $d_k \neq 0$, and so $\nabla f(x_k)^t d_k < 0$. This shows that d_k is a descent direction at x_k. Moreover, since $f(x)$ is a quadratic function, $f(x_k + \alpha d_k) = f(x_k) + \alpha \nabla f(x_k)^t d_k +$

146

$$\frac{1}{2}\alpha^2 d_k^t H d_k = f(x_k) + \alpha(\nabla f(x_k)^t d_k + \frac{1}{2} d_k^t H d_k) +$$

$\frac{1}{2}\alpha(\alpha - 1)d_k^t H d_k$, and therefore, we have from (6) that $f(x_k + \alpha d_k) < f(x_k)$ whenever $0 < \alpha \le 1$ and $d_k \neq 0$ solves QP(x_k). In the given algoirhtm, if a descent direction is found, then because we take $\alpha_k = 1$ if $x_k + d_k$ is feasible, or else, α_k computed by (5) then necessarily lies in the interval (0, 1), we have that $f(x_{k+1}) < f(x_k)$. The foregoing analysis asserts that the algorithm either verifies that a feasible solution is optimal, or else, through a finite number of working set reductions as in Part (b), it finds a $d_k \neq 0$ that yields a strictly better feasible solution as in Part (c).

Now, on the contrary, suppose that the algorithm does not terminate finitely and so generates an infinite number of iterations. Whenever $d_k = 0$ at any iteration, then either optimality is verified, or else, after a finite number of working set reductions, the algorithm either stops with an optimum or achieves a strict descent. Once the latter event occurs, the same working set cannot occur with (11.74) while yielding $d_k = 0$, since this would mean (by the uniqueness of the solution to (11.74)) that a previous iterate has repeated, a contradiction to the strict descent property. Hence, since there are only a finite number of W-sets, and each combination of a W-set and the event $d_k = 0$ cannot recur after a strict descent step, we must have $d_k \neq 0$ in the infinite sequence after a finite number of iterations, with each iteration then producing a strict descent. Now, at each such iteration, either the optimum $x_k + d_k$ to (11.74) corresponding to a certain W_k-set is feasible to QP, or else, the algorithm proceeds through a sequence of adding indices to the W-set, which must finitely produce feasibility to QP with a different W-set because of the strict descent property. However, since there are only a finite possible number of such W-sets, this contradicts the generation of an infinite sequence. Thus, the algorithm is finitely convergent.

d.   The problem of Example 11.2.1 to be solved is as follows:

Minimize $-2x_1 - 6x_2 + x_1^2 - 2x_1 x_2 + 2x_2^2$

subject to $x_1 + x_2 \le 2$

$$-x_1 + 2x_2 \le 2$$
$$x_1 \ge 0, \ x_2 \ge 0.$$

Therefore, for the four inequalities above $(E = \phi, \ I = \{1, 2, 3, 4\})$, we have

$$A_1 = \begin{bmatrix} 1 \\ 1 \end{bmatrix}, \ A_2 = \begin{bmatrix} -1 \\ 2 \end{bmatrix}, \ A_3 = \begin{bmatrix} -1 \\ 0 \end{bmatrix}, \ A_4 = \begin{bmatrix} 0 \\ 1 \end{bmatrix}, \ c = \begin{bmatrix} -2 \\ -6 \end{bmatrix}, \ \text{and}$$

$$H = \begin{bmatrix} 2 & -2 \\ -2 & 4 \end{bmatrix}.$$

Below, we denote vectors x^k and d^k with superscripts k to distinguish these from their components.

Iteration 1:

$$x^1 = [0 \ 0]^t, \ W_1 = I_1 = \{3, \ 4\}.$$

QP(x^1):   Minimize $-2d_1 - 6d_2 + d_1^2 - 2d_1d_2 + 2d_2^2$

                   subject to $d_1 = 0, \ d_2 = 0.$

The optimal solution is $d^1 = [0 \ 0]^t$, where the KKT system for QP(x^1) at d^1 gives $v_3 = -2$ and $v_4 = -6$. Hence, we drop the Constraint $i = 4$ from the working set. This gives

$$W_2 = I_2 = \{3\}, \ x^2 = x^1 = [0 \ 0]^t.$$

Iteration 2:

$$x^2 = [0 \ 0]^t, \ W_2 = I_2 = \{3\}.$$

QP(x^2):   Minimize $-2d_1 - 6d_2 + 2d_1^2 - d_1d_2 + 2d_2^2$

                   subject to $d_1 = 0.$

The optimal solution is $d^2 = [0 \ 3/2]^t$. The solution $x^2 + d^2$ violates Constraint $i = 2$, and so we need to determine a step length $\alpha_2 < 1$ as follows:

For Constraint $i = 2$: $A_2^t d^2 = 3$ and $b_2 - A_2^t x^2 = 2$.

Therefore, $\alpha_2 = 2/3$, and $q = 2$. This yields $W_3 = I_3 = \{2, 3\}$, and $x^3 = x^2 + \alpha_2 d^2 = [0 \ 0]^t$.

Iteration 3:

$x^3 = [0 \ 1]^t$, $W_3 = I_3 = \{2, 3\}$.

$QP(x^3):$ Minimize $-4d_1 - 2d_2 + d_1^2 - 2d_1 d_2 + 2d_2^2$
            subject to $-d_1 + 2 d_2 = 0$
                       $d_1 = 0.$

The optimal solution is $d^3 = [0 \ 0]^t$, where the KKT system for $QP(x^3)$ at d^3 gives $v_2 = 1$ and $v_3 = -5$. Hence, we drop the Constraint $i = 3$ from the working set. This yields

$W_4 = I_4 = \{2\}$, $x^4 = x^3 = [0 \ 1]^t$.

Iteration 4:

$x^4 = [0 \ 1]^t$, $W_4 = I_4 = \{2\}$.

$QP(x^4):$ Minimize $-4d_1 - 2d_2 + d_1^2 - 2d_1 d_2 + 2d_2^2$
            subject to $-d_1 + 2d_2 = 0.$

The optimal solution is $d^4 = [5 \ 5/2]^t$. The solution $x^4 + d^4$ violates Constraint $i = 1$, only, and so the step length α_4 is computed as

$$\alpha_4 = \frac{b_1 - A_1^t x^4}{A_1^t d^4} = \frac{1}{7.5} = \frac{2}{15},$$

149

and $q = 1$. This yields $W_5 = I_5 = \{1, 2\}$, and $x^5 = x^4 + \alpha_4 d^4 = [2/3 \ 4/3]^t$.

Iteration 5:

$x^5 = [2/3 \ 4/3]^t$, $W_5 = I_5 = \{1, 2\}$.

$QP(x^5):$ Minimize $-\dfrac{10}{3}d_1 - 2d_2 + d_1^2 - 2d_1 d_2 + 2d_2^2$

subject to $d_1 + d_2 = 0$

$-d_1 + 2d_2 = 0.$

The optimal solution is $d^5 = [0 \ 0]^t$, where the KKT system for $QP(x^5)$ at d^5 gives $v_1 = 25/9$ and $v_2 = -4/9$. Hence, we drop Constraint $i = 2$ from the working set, which yields

$W_6 = I_6 = \{1\}$, $x^6 = x^5 = [2/3 \ 4/3]^t$.

Iteration 6:

$x^6 = [2/3 \ 4/3]^t$, $W_6 = I_6 = \{1\}$.

$QP(x^6):$ Minimize $-\dfrac{10}{3}d_1 - 2d_2 + d_1^2 - 2d_1 d_2 + 2d_2^2$

subject to $d_1 + d_2 = 0.$

The optimal solution is $d^6 = [2/5 \ -2/15]^t$. The solution $x^6 + d^6 = [4/5 \ 6/5]^t$ satisfies all the constraints of QP, and so we take $x^7 = [4/5 \ 6/5]^t$, with $W_7 = I_7 = \{1\}$.

Iteration 7:

$x^7 = [4/5 \ 6/5]^t$, $W_7 = I_7 = \{1\}$.

$QP(x^7):$ Minimize $-\dfrac{14}{5}d_1 - \dfrac{14}{5}d_2 + d_1^2 - 2d_1 d_2 + 2d_2^2$

$$\text{subject to} \quad d_1 + d_2 = 0.$$

The optimal solution is $d^7 = [0 \; 0]^t$, where the KKT system for QP(x^7) at d^7 gives $v_1 = 14/5$. We therefore stop with $x^* = [4/5 \; 6/5]^t$ as an optimal solution to Problem QP.

11.22 The matrix H is positive semidefinite, and so $f(x) = c^t x + \frac{1}{2} x^t H x$ is convex. Since f is also differentiable, this implies that $f(x)$ is unbounded from below in R^n if and only if it does not have a minimizing solution, which happens if and only if no vector x exists for which $\nabla f(x) = 0$. Since $\nabla f(x) = c + Hx$, the proof is complete. □

11.23 a. Suppose that $G_r x \leq g_r$ is implied by the remaining inequalities $G_i x \leq g_i$ for $i = 1, \ldots, \bar{m}$, $i \neq r$. Let the latter constraints be denoted by $\bar{G}x \leq \bar{g}$. Then we have by LP duality that

$$g_r \geq \max\{G_r x : \bar{G}x \leq \bar{g}\} = \min\{\bar{g}^t u : u^t \bar{G} = G_r, \; u \geq 0\}.$$

Hence, since an optimum exists (given feasibility of $\bar{G}x \leq \bar{g}$ and that $g_r < \infty$), we have that

there exists $\bar{u} \geq 0 : \bar{u}^t \bar{G} = G_r$ and $\bar{g}^t \bar{u} \leq g_r$. \hfill (1)

Now, for any $k \in \{1, \ldots, \bar{m}\}/r$, consider the RLT constraint

$$[(g_r - G_r x)(g_k - G_k x)]_L \geq 0 \qquad (2)$$

where $[\cdot]_L$ denotes the linearized version of the expression in $[\cdot]$ under the substitution (11.16). We need to show that (2) is implied by the remaining RLT inequalities, which in particular, include the following:

$$[(g_k - G_k x)(\bar{g} - \bar{G}x)]_L \geq 0. \qquad (3)$$

Taking the inner product of (3) with $\bar{u} \geq 0$ as given by (1) yields

$$[(g_k - G_k x)(\bar{u}^t \bar{g} - G_r x)]_L \geq 0,$$

i.e., $(g_k - G_k x)\bar{u}^t \bar{g} - [(g_k - G_k x)(G_r x)]_L \geq 0.$ $\qquad$ (4)

But $\bar{u}^t \bar{g} \leq g_r$ by (1) and $(g_k - G_k x) \geq 0$ by feasibility to the RLT system, and so (4) implies that

$$(g_k - G_k x)g_r - [(g_k - G_k x)(G_r x)]_L \geq 0,$$

i.e., $[(g_k - G_k x)(g_r - G_r x)]_L \geq 0,$ or that (2) holds true. $\quad\square$

b. For Example 11.28, note that the weighted sum of the defining inequalities

$$\{\frac{1}{8}(120 - 3x_1 - 8x_2 \geq 0) + \frac{3}{8}(x_1 \geq 0)\} \text{ yields } (15 - x_2) \geq 0 \qquad (5)$$

and so $x_2 \leq 15$ is implied by $3x_1 + 8x_2 \leq 120$ and $x_1 \geq 0$. Accordingly, in the notation of Part (a), any RLT inequality of the type

$$[(15 - x_2)(g_k - G_k x)]_L \geq 0 \qquad (6)$$

is implied by the remaining RLT inequalities, which, in particular, include the following RLT restrictions:

$$[(120 - 3x_1 - 8x_2)(g_k - G_k x)]_L \geq 0 \qquad (7a)$$

$$[x_1(g_k - G_k x)]_L \geq 0, \qquad (7b)$$

as seen by weighting (7a) and (7b) by 1/8 and 3/8, respectively, as in (5), and summing to get (6).

The 15 RLT inequalities obtained by omitting $(15 - x_2) \geq 0$ from the pairwise products are given by the constraints of LP(Ω^1) in the solution to Exercise 11.24 below, where this LP yields the same optimal value (and solution) as does LP(Ω) in Example 11.2.8.

11.24 Consider LP(Ω^1). Since $x_2 \leq 15$ is implied by the second structural constraint, we drop this from the formulation and we consider the following bound-and-constraint-factors:

$$x_1 \geq 0$$
$$x_2 \geq 0$$
$$24 - x_1 \geq 0$$
$$24 + 3x_1 - 4x_2 \geq 0$$
$$120 - 3x_1 - 8x_2 \geq 0.$$

Taking pairwise products of these factors (systematically) including self-products produces $\binom{5+1}{2} = \binom{6}{2} = 15$ constraints to yield the following formulation (note that the above restrictions are all implied by the constraints of LP(Ω^1)):

LP(Ω^1) : Minimize $\quad -w_{11} - w_{22} + 24x_1 - 144$

subject to

$$w_{11} \geq 0$$
$$w_{12} \geq 0$$
$$24x_1 - w_{11} \geq 0$$
$$24x_1 + 3w_{11} - 4w_{12} \geq 0$$
$$120x_1 - 3w_{11} - 8w_{12} \geq 0$$
$$w_{22} \geq 0$$
$$24x_2 - w_{12} \geq 0$$
$$24x_2 + 3w_{12} - 4w_{22} \geq 0$$
$$120x_2 - 3w_{12} - 8w_{22} \geq 0$$
$$576 + w_{11} - 48x_1 \geq 0$$
$$576 + 48x_1 - 96x_2 - 3w_{11} + 4w_{12} \geq 0$$
$$2880 - 192x_1 - 192x_2 + 3w_{11} + 8w_{12} \geq 0$$
$$576 + 9w_{11} + 16w_{22} + 144x_1 - 192x_2 - 24w_{12} \geq 0$$
$$2880 + 288x_1 - 672x_2 - 9w_{11} - 12w_{12} + 32w_{22} \geq 0$$
$$14400 + 9w_{11} + 64w_{22} - 720x_1 - 1920x_2 + 48w_{12} \geq 0.$$

The optimal solution is given by $(\overline{x}_1, \overline{x}_2, \overline{w}_{11}, \overline{w}_{12}, \overline{w}_{22}) = (8, 6, 192,$ 48, 72), with $v[LP(\Omega^1)] = -216$. The solution $(\overline{x}_1, \overline{x}_2) = (8, 6)$ is a feasible solution to the original problem NQP with objective value -52. Hence currently, $(x_1^*, x_2^*) = (8, 6)$, with $v^* = -52$ and $LB = -216$. As in Example 11.2.8, we branch on x_1 at the value $8 \in (0, 24)$ to obtain the following partitioning:

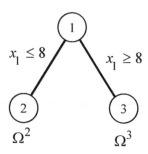

Formulation of $LP(\Omega^2)$:

Consider the following bound-and constraint-factor restrictions:

$$x_1 \geq 0 \tag{1a}$$
$$x_2 \geq 0 \tag{1b}$$
$$8 - x_1 \geq 0 \tag{1c}$$
$$24 + 3x_1 - 4x_2 \geq 0 \tag{1d}$$

Note that (1d) and (1c) imply that $x_2 \leq 12$ (hence, $x_2 \leq 15$ is implied), and that (1c) and $x_2 \leq 12$ imply that $3x_1 + 8x_2 \leq 24 + 96 = 120$. Hence for $LP(\Omega^2)$, we only need to consider the $\begin{pmatrix} 4+1 \\ 2 \end{pmatrix} = 10$ pairwise products of (1a – d), including self-products. This yields the following formulation:

$LP(\Omega^2):$ Minimize $-w_{11} - w_{22} + 24x_1 - 144$

subject to $w_{11} \geq 0$

$w_{12} \geq 0$

$8x_1 - w_{11} \geq 0$

154

$$24x_1 + 3w_{11} - 4w_{12} \geq 0$$
$$w_{22} \geq 0$$
$$8x_2 - w_{12} \geq 0$$
$$24x_2 + 3w_{12} - 4w_{22} \geq 0$$
$$64 + w_{11} - 16x_1 \geq 0$$
$$192 - 32x_2 - 3w_{11} + 4w_{12} \geq 0$$
$$576 + 9w_{11} + 16w_{22} + 144x_1 - 192x_2 - 24w_{12} \geq 0.$$

The optimal solution to $LP(\Omega^2)$ is given by $(\bar{x}_1, \bar{x}_2, \bar{w}_{11}, \bar{w}_{12}, \bar{w}_{22}) =$ $(0, 6, 0, 0, 36)$ with $v[LP(\Omega^2)] = -180$. The solution $(\bar{x}_1, \bar{x}_2) = (0, 6)$ is feasible to NPQ with objective value -180. Hence, since $-180 < -52$, we update $(x_1^*, x_2^*) = (0, 6)$ and $v^* = -180$, and we fathom Node 2.

Formulation of $LP(\Omega^3)$:

Consider the following bound-and constraint-factors:

$$x_1 - 8 \geq 0 \tag{2a}$$
$$x_2 \geq 0 \tag{2b}$$
$$24 - x_1 \geq 0 \tag{2c}$$
$$24 + 3x_1 - 4x_2 \geq 0 \tag{2d}$$
$$120 - 3x_1 - 8x_2 \geq 0. \tag{2e}$$

Note that these include all the original restrictions of the problem except for $x_1 \geq 0$, which is implied by (2a), and $x_2 \leq 15$, which is implied by (2a) and (2e), where the latter actually yield $x_2 \leq 12$. Hence, taking pairwise products of (2a) – (2e), including self-products, produces the following model with $\binom{6}{2} = 15$ constraints:

$LP(\Omega^3)$: Minimize $-w_{11} - w_{22} + 24x_1 - 144$
   subject to $64 + w_{11} - 16x_1 \geq 0$
   $w_{12} - 8x_2 \geq 0$
   $32x_1 - w_{11} - 192 \geq 0$

$$3w_{11} - 4w_{12} + 32x_2 - 192 \geq 0$$
$$144x_1 + 64x_2 - 3w_{11} - 8w_{12} - 960 \geq 0$$
$$w_{22} \geq 0$$
$$24x_2 - w_{12} \geq 0$$
$$24x_2 + 3w_{12} - 4w_{22} \geq 0$$
$$120x_2 - 3w_{12} - 8w_{22} \geq 0$$
$$576 + w_{11} - 48x_1 \geq 0$$
$$576 + 48x_1 - 96x_2 - 3w_{11} + 4w_{12} \geq 0$$
$$2880 - 192x_1 - 192x_2 + 3w_{11} + 8w_{12} \geq 0$$
$$576 + 9w_{11} + 16w_{22} + 144x_1 - 192x_2 - 24w_{12} \geq 0$$
$$2880 + 288x_1 - 672x_2 - 9w_{11} - 12w_{12} + 32w_{22} \geq 0$$
$$14400 + 9w_{11} + 64w_{22} - 720x_1 - 1920x_2 + 48w_{12} \geq 0.$$

The optimal solution to $\text{LP}(\Omega^3)$ is given by $(\bar{x}_1, \bar{x}_2, \bar{w}_{11}, \bar{w}_{12}, \bar{w}_{22}) =$ (24, 6, 576, 144, 36) with objective value $v[\text{LP}(\Omega^3)] = -180$. Also, the feasible solution $(\bar{x}_1, \bar{x}_2) = (24, 6)$ yields an objective value of -180 in Problem NPQ. Hence, we fathom Node 3 as well, and the solutions (x_1^*, x_2^*) equal to $(0, 6)$ or $(24, 6)$ are alternative optimal solutions to Problem NPQ of objective value -180.

11.36 By substituting $\Delta_j = \mu_{j+1} - \mu_j$ as given in Exercise 11.35 in the δ-form: we obtain the following representation for x:

$$x = \mu_1 + \sum_{j=1}^{k-1}(\mu_{j+1} - \mu_j)\delta_j = \mu_1 + \sum_{j=2}^{k-1}\mu_j\delta_{j-1} + \mu_k\delta_{k-1} - \sum_{j=2}^{k-1}\mu_j\delta_j - \mu_1\delta_1$$

$$= \mu_1(1 - \delta_1) + \sum_{j=2}^{k-1}\mu_j(\delta_{j-1} - \delta_j) + \mu_k\delta_{k-1}.$$

Hence,

$$x = \sum_{j=1}^{k-1}\mu_j(\delta_{j-1} - \delta_j) + \mu_k\delta_{k-1}, \tag{1}$$

where $\delta_0 \equiv 1$. Therefore, the equations representing x in the λ-form and in the δ-form are equivalent if λ_j, $j = 1,.., k$, is as given in the exercise.

Next, define a $k \times k$ matrix $T = [t_{ij}]$, where $t_{ii} = 1$ for $i = 1,..., k$, $t_{i,i+1} = -1$ for $i = 1,..., k-1$, and $t_{ij} = 0$ otherwise. Then, $\lambda = T\delta$ where $\lambda \equiv (\lambda_1,...\lambda_k)^t$ and $\delta \equiv (\delta_0, \delta_1,..., \delta_{k-1})^t$. The matrix T is upper triangular with ones along the diagonal, and is therefore invertible.

To see how the other restrictions in λ and δ are related via this relationship, note that because $\delta_0 = 1$, $0 \le \delta_j \le 1$ for $j = 1,..., k-1$, and because $\delta_i = 0$ implies that $\delta_j = 0$ for all $j > i$, we have that all elements of the vector λ computed via $\lambda = T\delta$ lie in the interval $[0, 1]$. Moreover, $e^t\lambda = e^t T\delta = \delta_0 = 1$, where e is a conformable vector of ones. Also, since $\delta = T^{-1}\lambda$, where T^{-1} is an upper triangular matrix whose diagonal entries and all entries above it are equal to 1, we obtain $\delta_0 = 1$ and $0 \le \delta_j \le 1$ for $j = 1,..., k-1$ whenever $\lambda_j \ge 0$ for $j = 1,...,$ k and $\sum_{j=1}^{k} \lambda_j = 1$. What remains to be verified is that the two nonlinear requirements are also equivalent.

Consider the requirement "$\lambda_p \lambda_q = 0$ if μ_p and μ_q are not adjacent," and suppose that for some $p \in \{1,..., k-1\}$ we have $\lambda_p > 0$ and $\lambda_{p+1} > 0$. Then $\lambda_j = 0$ for the remaining indices j. From the relationship between λ and δ (in particular, using $\delta = T^{-1}\lambda$) we then obtain

$$\delta_j = 1 \text{ for } j = 0,1,..., p-1$$
$$\delta_p = \lambda_{p+1} = 1 - \lambda_p, \text{ and}$$
$$\delta_{p+r} = 0 \text{ for } r = 1,..., k-p.$$

Also, if $\lambda_p = 1$ for some $p \in \{1,..., k\}$ and $\lambda_j = 0$ for all $j \ne p$, then $\delta_j = 1$ for $j \le p$, and $\delta_j = 0$ for $j > p$. Thus the requirement "$\delta_i > 0$ implies that $\delta_j = 1$ for $j < i$" is met.

It can be shown similarly by viewing the form of $\lambda = T\delta$ that if δ_j, $j = 0, 1,..., k-1$, satisfy the restriction that "$\delta_i > 0$ implies that $\delta_j = 1$ for j

$< i$," then $\lambda_j, j = 1,\ldots,k$, are such that " $\lambda_p \lambda_q = 0$ if μ_p and μ_q are not adjacent." Therefore, the two requirements are equivalent. This completes the proof of the equivalence of the two representations. $\qquad\square$

11.41 In the case of linear fractional programs, if the feasible region is not bounded, then an optimal solution may not exist. This does not necessarily mean that the objective function (to be minimized) is unbounded below. As opposed to linear programs we may be faced with the case where the objective function is bounded but does not attain its lower bound on the feasible region.

In particular, consider the line search along the direction identified in the exercise. Without loss of generality, suppose that the objective function $f(x) = (p^t x + \alpha)/(q^t x + \beta)$ satisfies $q^t x + \beta > 0$ for all $x \in X \equiv \{x : Ax = b, \ x \geq 0\}$. Consider the solution $\bar{x}$ at which the search direction $d = \begin{bmatrix} d_B \\ d_N \end{bmatrix}$ satisfies $d_N = e_j$ and $d_B = -B^{-1} a_j \geq 0$, where e_j is the jth unit vector, so that d is a recession direction of X. We then have

$$\theta(\lambda) \equiv f(\bar{x} + \lambda d) = \frac{p_B^t(\bar{x}_B + \lambda d_B) + p_N^t(\bar{x}_N + \lambda d_N) + \alpha}{q_B^t(\bar{x}_B + \lambda d_B) + q_N^t(\bar{x}_N + \lambda d_N) + \beta}$$

$$= \frac{p_B^t \bar{x}_B + p_N^t \bar{x}_N + \alpha + \lambda[p_{Nj} - p_B^t B^{-1} a_j]}{q_B^t \bar{x}_B + q_N^t \bar{x}_N + \beta + \lambda[q_{Nj} - q_B^t B^{-1} a_j]},$$

i.e., in obvious notation, $\theta(\lambda)$ is of the form

$$\theta(\lambda) \equiv f(\bar{x} + \lambda d) = \frac{p_0 + \lambda \bar{p}_j}{q_0 + \lambda \bar{q}_j}, \quad \text{where } \bar{q}_j \geq 0 \qquad (1)$$

since $q_0 + \lambda \bar{q}_j > 0, \ \forall \lambda \geq 0$. Note also from (1) that

$$\theta'(\lambda) = \frac{q_0 \bar{p}_j - p_0 \bar{q}_j}{[q_0 + \lambda \bar{q}_j]^2} < 0, \quad \forall \lambda \geq 0 \text{ by Lemma 11.4.2.} \qquad (2)$$

Now, consider the following two cases:

158

Case (i): $\bar{q}_j = 0$. In this case, since $q_0 = q^t \bar{x} + \beta > 0$, we get from (2) that $\bar{p}_j < 0$, and so this implies from (1) that $\theta(\lambda) \to -\infty$ as $\lambda \to \infty$, and the objective value is indeed unbounded by moving from $\bar{x}$ along the direction d.

Case (ii): $\bar{q}_j > 0$. In this case, $\theta(\lambda) \to \bar{p}_j / \bar{q}_j$, where from (2), we get $-\infty < \bar{p}_j / \bar{q}_j < p_0 / q_0 = f(\bar{x})$. Hence, in this case, the objective value decreases toward the finite lower bounding value $\bar{p}_j / \bar{q}_j$ as $\lambda \to \infty$.

11.42 a. For any λ_1 and $\lambda_2 \in R$, and for any $\alpha \in [0, 1]$ we have

$$\theta(\alpha\lambda_1 + (1 - \alpha)\lambda_2) = f[x + (\alpha\lambda_1 + (1 - \alpha)\lambda_2)d]$$
$$= f[\alpha x + (1 - \alpha)x + (\alpha\lambda_1 + (1 - \alpha)\lambda_2)d]$$
$$= f[\alpha(x + \lambda_1 d) + (1 - \alpha)(x + \lambda_2 d)]$$
$$\geq \min\{f(x + \lambda_1 d),\ f(x + \lambda_2 d)\} = \min\{\theta(\lambda_1),\ \theta(\lambda_2)\}.$$

Here, the above inequality follows from the assumed quasiconcavity of the function $f(x)$. The foregoing derivation shows that the function $\theta(\lambda)$ is also quasiconcave.

b. From Theorem 3.5.3, we can conclude that the minimum of the function $f(x + \lambda d)$ over an interval $[0, b]$ must occur at one of the two endpoints. However, at $\lambda = 0$ we have $\theta'(0) = \nabla f(x)^t d < 0$, and so the minimum value must occur for $\lambda = b$.

c. By Lemma 11.4.1, the given fractional function is quasiconcave, and so from Part (b), in the case of the convex simplex method, the linear search process reduces to evaluating λ_{max}, and then directly setting λ_k equal to λ_{max}. (Also, see Lemma 11.4.2 for a related argument.)

11.47 By defining $x_0 \equiv f_2(x)$, we obtain the following equivalent problem:

$$\text{Minimize} \quad f_1(x) + x_0^a f_3(x)$$
$$\text{subject to} \quad x_0 = f_2(x).$$

159

However, because the functions $f_2(x)$ and $f_3(x)$ take on positive values, and the objective function is to be minimized, we can replace the equality constraint with the inequality $x_0 \geq f_2(x)$, since this constraint will automatically be satisfied as an equality at optimality. Furthermore, since $f_2(x)$ is positive for any positive x, so is x_0. This allows us to rewrite the problem as follows:

$$\text{Minimize}\{ f_1(x) + x_0^a f_3(x) : x_0^{-1} f_2(x) \leq 1, \ x > 0\}. \tag{1}$$

Finally, note that if a function $h(x)$ is a posynomial in $x_1, \ldots, x_n$, and if $x_0 > 0$, then for any real a, $x_0^a h(x)$ is a posynomial in $x_0, x_1, \ldots, x_n$. Also, a sum of posynomial functions is a posynomial. Therefore, the problem given by (1) is in the form of the standard posynomial geometric program.

The numerical example can be restated as follows:

$$\text{Minimize} \quad 2x_1^{-1/3} x_2^{1/6} + x_0^{1/2} x_1^{3/4} x_2^{-1/3}$$

$$\text{subject to} \quad \frac{3}{5} x_0^{-1} x_1^{1/2} x_2^{3/4} + \frac{2}{5} x_0^{-1} x_1^{2/3} x_2 \leq 1$$

$$x > 0.$$

Hence, we have:

$M = 4$, $J_0 = \{1, 2\}$, $J_1 = \{3, 4\}$, $n = 3$ (thus $DD = 0$), $\alpha_1 = 2$, $\alpha_2 = 1$, $\alpha_3 = 3/5$, $\alpha_4 = 2/5$, $a_1^t = [0 \ -1/3 \ 1/6]$, $a_2^t = [1/2 \ 3/4 \ -1/3]$, $a_3^t = [-1 \ 1/2 \ 3/4]$, and $a_4^t = [-1 \ 2/3 \ 1]$.

Step 1. Solve the dual problem:

Since $DD = 0$, the dual problem has a unique feasible solution as given by the following constraints of Problem DGP in terms of δ_i, $i = 1, \ldots, 4$, and u_1:

$$\frac{1}{2}\delta_2 \quad - \quad \delta_3 \quad - \quad \delta_4 \quad = 0$$

$$-\frac{1}{3}\delta_1 \quad + \quad \frac{3}{4}\delta_2 \quad + \quad \frac{1}{2}\delta_3 \quad + \quad \frac{2}{3}\delta_4 \quad = 0$$

$$\frac{1}{6}\delta_1 \quad - \quad \frac{1}{3}\delta_2 \quad + \quad \frac{3}{4}\delta_3 \quad + \quad \delta_4 \quad = 0$$

$$\delta_1 \quad + \quad \delta_2 \qquad\qquad\qquad = 1$$

$$\qquad\qquad\qquad \delta_3 \quad + \quad \delta_4 -u_1 = 0.$$

The unique solution to this system, and therefore the unique optimal solution to the dual problem, is given by

$$\delta_1 = \frac{35}{51}, \; \delta_2 = \frac{16}{51}, \; \delta_3 = \frac{34}{51}, \; \delta_4 = \frac{-26}{51}, \text{ and } u_1 = \frac{8}{51}. \tag{2}$$

Note that $\delta_4 < 0$ and so the dual is infeasible. In fact, there does not exist a KKT solution for (11.53). To see this, note that (11.53) is given by

$$\text{Minimize } \ell n[\tau_1 + \tau_2]$$
$$\text{subject to } \ell n[\tau_3 + \tau_4] \le 0.$$

Denoting u_1 as the Lagrange multiplier, the KKT system is given by

$$\frac{\tau_1 a_1 + \tau_2 a_2}{(\tau_1 + \tau_2)} + \frac{u_1[\tau_3 a_3 + \tau_4 a_4]}{(\tau_3 + \tau_4)} = 0 \tag{3}$$

$$u_1 \ge 0, \; (\tau_3 + \tau_4) \le 1, \; u_1[\tau_3 + \tau_4 - 1] = 0. \tag{4}$$

Denoting $\delta_1 = \dfrac{\tau_1}{\tau_1 + \tau_2}$, $\delta_2 = \dfrac{\tau_2}{\tau_1 + \tau_2}$, $\delta_3 = \dfrac{u_1 \tau_3}{\tau_3 + \tau_4}$, and $\delta_4 = \dfrac{u_1 \tau_4}{\tau_3 + \tau_4}$

as in (11.57), we get from (3) that $\sum\limits_{k=1}^{4} \delta_k a_k = 0$, where $\delta_1 + \delta_2 = 1$, and $u_1 = \delta_3 + \delta_4$. But this leads to the unique solution (2) where $\delta_4 < 0$, a contradiction. (Note that with $u_1 = 0$, (3) yields $a_1 \delta_1 + a_2 \delta_2 = 0$, which yields $\delta_1 = \delta_2 = 0$, contradicting $\delta_1 + \delta_2 = 1$.) Hence, no KKT solution exists.

To provide additional insight, note that if we consider the solution $x_1 = 1/\lambda^{1/25}$ and $x_2 = \lambda$, for $\lambda > 0$, with $x_0 = \frac{3}{5}x_1^{1/2}x_2^{3/4} + \frac{2}{5}x_1^{2/3}x_2 = \frac{3}{5}\lambda^{73/100} + \frac{2}{5}\lambda^{73/75}$, then this defines a feasible trajectory with objective value $2\lambda^{27/150} + \left[\dfrac{3}{5}\lambda^{1/300} + \dfrac{2}{5}\lambda^{37/150}\right]^{1/2}$. Hence, as $\lambda \to 0^+$, the objective value approaches zero, which is a lower bound for the posynomial, with $x_1 \to \infty$ and $x_2 \to 0^+$. Hence, an optimum does not exist, although zero is the infimum value.

11.48 For the given problem, we have $M = 5$ terms and $n = 3$ variables. Hence, the degree of difficulty DD $= 1$. To formulate the dual program, DGP, note that

$$J_0 = \{1,\ 2,\ 3\} \text{ and } J_1 = \{4,\ 5\} \text{ with } m = 1;\ \alpha_1 = 25,\ \alpha_2 = 20,\ \alpha_3 = 30,\ \alpha_4 = \dfrac{5}{3}, \text{ and } \alpha_5 = \dfrac{4}{3};$$

$$a_1 = \begin{bmatrix} -2 \\ -1/2 \\ -1 \end{bmatrix},\ a_2 = \begin{bmatrix} 2 \\ 0 \\ 1 \end{bmatrix},\ a_3 = \begin{bmatrix} 1 \\ 2 \\ 1 \end{bmatrix},\ a_4 = \begin{bmatrix} -1 \\ -2 \\ 0 \end{bmatrix}, \text{ and } a_5 = \begin{bmatrix} 0 \\ 1/2 \\ -2 \end{bmatrix}.$$

This yields the following dual program:

DGP: Maximize $\displaystyle\sum_{k=1}^{5} \delta_k \ell n[\alpha_k / \delta_k] + u_1 \ell n(u_1)$ (1)

subject to

$$-2\delta_1 + 2\delta_2 + \delta_3 - \delta_4 = 0 \tag{2}$$

$$-\dfrac{1}{2}\delta_1 + 2\delta_3 - 2\delta_4 + \dfrac{1}{2}\delta_5 = 0 \tag{3}$$

$$-\delta_1 + \delta_2 + \delta_3 - 2\delta_5 = 0 \tag{4}$$

$$\delta_1 + \delta_2 + \delta_3 = 1 \tag{5}$$

$$\delta_4 + \delta_5 = u_1 \tag{6}$$

$$(\delta,\ u_1) \ge 0. \tag{7}$$

From Equations (2) – (6), we get that

$$\delta_2 = \dfrac{3}{4}\delta_1 + \dfrac{1}{16},\ \delta_3 = \dfrac{15}{16} - \dfrac{7}{4}\delta_1,\ \delta_4 = \dfrac{17}{16} - \dfrac{9}{4}\delta_1,\ \delta_5 = \dfrac{1}{2} - \delta_1, \text{ and}$$

$$u_1 = \dfrac{25}{16} - \dfrac{13}{14}\delta_1. \tag{8}$$

The restrictions $(\delta,\ u_1) \ge 0$ along with (8) yield

162

$$0 \le \delta_1 \le \frac{17}{36}. \tag{9}$$

Projecting Problem DGP onto the space of δ_1 by using (8) and (9) and solving this resultant problem yields the following solution (δ^*, u_1^*) (upon using (8) and (9)):

$$\delta_1^* = 0.06967, \ \delta_2^* = 0.11475, \ \delta_3^* = 0.81558, \ \delta_4^* = 0.90575, \ \delta_5^* =$$
$$0.43033, \text{ and } u_1^* = 1.33608, \tag{10a}$$

with objective value

$$v^* = 5.36836. \tag{10b}$$

Using Equations (11.71a, b), we therefore get

$$
\begin{aligned}
-2y_1 - 0.5y_2 - y_3 &= -0.5145 \\
2y_1 \qquad\quad +y_3 &= 0.20763 \\
y_1 +2y_2 + y_3 &= 1.76331 \\
-y_1 -2y_2 \qquad &= -0.89956 \\
0.5y_2 - 2y_3 &= -1.42062.
\end{aligned}
$$

The solution to the above (consistent) system is given by

$$y_1^* = -0.32792, \ y_2^* = 0.61374, \text{ and } y_3^* = 0.86347. \tag{11}$$

Using the fact that $x_j = e^{y_j}$, we obtain the following optimal solution to the original problem:

$x_1^* = 0.72042, \ x_2^* = 1.84733, \text{ and } x_3^* = 2.37138$, with optimal objective value $e^{v^*} = 214.51078$ as given via (10b) (since $v^* = \ell n[F(y^*)]$).

11.50 The given problem can be formulated as follows:

Minimize $\quad 2\sqrt{x_1^2 + x_2^2} + 6x_1 + 6x_2 + 4x_3$

subject to

163

$$x_1 x_2 x_3 \geq 15$$
$$(x_1,\ x_2,\ x_3) > 0.$$

Letting $x_0 \equiv x_1^2 + x_2^2$, this problem can be equivalently posed (see Exercise 11.47) as the following posynomial geometric program (GP):

GP: Minimize $2x_0^{1/2} + 6x_1 + 6x_2 + 4x_3$

subject to
$$15x_1^{-1}x_2^{-1}x_3^{-1} \leq 1$$
$$x_0^{-1}x_1^2 + x_0^{-1}x_2^2 \leq 1$$
$$(x_0,\ x_1,\ x_2,\ x_3) > 0.$$

For Problem GP, we have $M = 7$, $n = 4$, $DD = M - n - 1 = 2$, $m = 2$, $J_0 = \{1, 2, 3, 4\}$, $J_1 = \{5\}$, $J_2 = \{6, 7\}$, with $\alpha_1 = 2$, $\alpha_2 = 6$, $\alpha_3 = 6$, $\alpha_4 = 4$, $\alpha_5 = 15$, $\alpha_6 = 1$, $\alpha_7 = 1$, and with

$$a_1 = \begin{bmatrix} 1/2 \\ 0 \\ 0 \\ 0 \end{bmatrix},\ a_2 = \begin{bmatrix} 0 \\ 1 \\ 0 \\ 0 \end{bmatrix},\ a_3 = \begin{bmatrix} 0 \\ 0 \\ 1 \\ 0 \end{bmatrix},\ a_4 = \begin{bmatrix} 0 \\ 0 \\ 0 \\ 1 \end{bmatrix},\ a_5 = \begin{bmatrix} 0 \\ -1 \\ -1 \\ -1 \end{bmatrix},$$

$$a_6 = \begin{bmatrix} -1 \\ 2 \\ 0 \\ 0 \end{bmatrix},\ \text{and } a_7 = \begin{bmatrix} -1 \\ 0 \\ 2 \\ 0 \end{bmatrix}.$$

Hence, the dual geometric program is given as follows:

DGP: Maximize $\sum\limits_{k=1}^{7} \delta_k \ell n \left[\dfrac{\alpha_k}{\delta_k} \right] + u_1 \ell n(u_1) + u_2 \ell n(u_2)$

subject to
$$\frac{1}{2}\delta_1 - \delta_6 - \delta_7 = 0$$
$$\delta_2 - \delta_5 + 2\delta_6 = 0$$
$$\delta_3 - \delta_5 + 2\delta_7 = 0$$
$$\delta_4 - \delta_5 = 0$$
$$\delta_1 + \delta_2 + \delta_3 + \delta_4 = 1$$
$$u_1 = \delta_5$$

164

$$u_2 = \delta_6 + \delta_7$$
$$(\delta, u) \geq 0.$$

Projecting Problem DGP onto the space of (δ_1, δ_2), we get

$$\delta_3 = \frac{2}{3} - \delta_1 - \delta_2, \ \delta_4 = \frac{1}{3}, \ \delta_5 = \frac{1}{3}, \ \delta_6 = \frac{1}{6} - \frac{1}{2}\delta_2,$$

$$\delta_7 = -\frac{1}{6} + \frac{1}{2}\delta_1 + \frac{1}{2}\delta_2, \ u_1 = \frac{1}{3}, \text{ and } u_2 = \frac{1}{2}\delta_1. \tag{1}$$

Solving this projected problem and using (1) yields the following solution (δ^*, u^*):

$\delta_1^* = 0.12716, \ \delta_2^* = 0.26975, \ \delta_3^* = 0.26975, \ \delta_4^* = 0.33333, \ \delta_5^* = 0.33333, \ \delta_6^* = 0.03179, \ \delta_7^* = 0.03179, \ u_1^* = 0.33333, \text{ and } u_2^* = 0.06358,$

with objective value $v^* = 3.79899$. The system (11.71) for recovering the y-variables is given as follows:

$$\frac{1}{2}y_0 = 1.04353$$
$$y_1 = 0.69697$$
$$y_2 = 0.69697$$
$$y_3 = 1.31407$$
$$-y_1 - y_2 - y_3 = -2.70805$$
$$-y_0 + 2y_1 = -0.69315$$
$$-y_0 + 2y_2 = -0.69315.$$

This system yields (consistent up to four decimal places)

$$y_0^* = 2.08706, \ y_1^* = 0.69697, \ y_2^* = 0.69697, \text{ and } y_3^* = 1.31407.$$

Accordingly, using $x_j = e^{y_j}, \ \forall j = 0, 1, 2, 3$, we get

$$x_0^* = 8.06118, \ x_1^* = 2.00766, \ x_2^* = 2.00766, \text{ and } x_3^* = 3.72129, \text{ with}$$

objective value $e^{v^*} = 44.65606$ (since $v^* = \ell n[F(y^*)]$).

11.51 Let x^* solve the problem to minimize $f_1(x) - f_2(x)$. By assumption, $f_2(x^*) - f_1(x^*) > 0$. Therefore, (x_0^*, x^*) solves the following problem:

$$\text{Maximize } \{x_0 : x_0 \le f_2(x) - f_1(x)\}.$$

Furthermore, since $f_2(x)$ is a positive-valued function, and since the maximization of x_0 is equivalent here to minimizing its reciprocal, we obtain the following equivalent optimization problem:

$$\text{Minimize } \{x_0^{-1} : \frac{x_0}{f_2(x)} + \frac{f_1(x)}{f_2(x)} \le 1\}.$$

Finally, we note that by same arguments as those in the solution to Exercise 11.52 below, it can be easily shown that both the objective function and the function on the left-hand side of the constraint are posynomials. □

11.52 Throughout, we assume that $f_3(x) - f_4(x) > 0$, $\forall x > 0$, since a can be a general rational exponent. The problems

$$\textbf{P1:} \quad \text{Minimize} \quad f_1(x) + \frac{f_2(x)}{[f_3(x) - f_4(x)]^a}$$

and

$$\textbf{P2:} \quad \text{Minimize} \quad f_1(x) + f_2(x)x_0^{-a}$$

$$\text{subject to} \quad x_0 \le f_3(x) - f_4(x)$$

are clearly equivalent. Also, note that $f_3(x)$ is positive-valued, and therefore the constraint in Problem P2 can be rewritten as

$$\frac{x_0}{f_3(x)} + \frac{f_4(x)}{f_3(x)} \le 1. \tag{1}$$

It remains to show that the objective function as well as the expression on the left-hand side of (1) are posynomials.

Readily, if $f_1(x)$ and $f_2(x)$ are posynomials in $x_1,...,x_n$, then $f_1(x) + f_2(x)x_0^{-a}$ is a posynomial in x_0, $x_1,...,x_n$. By assumption, $f_3(x)$ is a single-term posynomial, say, of the form $f_3(x) = \alpha \prod_{j=1}^{n} x_j^{a_j}$, where $\alpha > 0$ and a_j, $j = 1,...,n$, are rational exponents. Then

$$\frac{x_0}{f_3(x)} = \frac{1}{\alpha} \prod_{j=0}^{n} x_j^{\bar{a}_j}, \text{ where } \bar{a}_j \equiv -a_j \text{ for } j = 1,...,n, \text{ and } \bar{a}_0 \equiv 1. \text{ Hence,}$$

$$\frac{x_0}{f_3(x)} \text{ is a posynomial. Similarly, in the notation of (11.49), let}$$

$f_4(x) = \sum_{k \in J_4} \alpha_k \prod_{j=1}^{n} x_j^{a_{kj}}$. Then $\dfrac{f_4(x)}{f_3(x)} = \sum_{k \in J_4} \bar{\alpha}_k \prod_{j=1}^{n} x_j^{\bar{a}_{kj}}$, where for each $k \in J_4$, we have $\bar{\alpha}_k \equiv \alpha_k / \alpha$ and $\bar{a}_{kj} = a_{kj} - a_j$. This shows that the constraint function is also a posynomial, and this completes the proof. $\square$

Printed and bound by CPI Group (UK) Ltd, Croydon, CR0 4YY

27/10/2024

14580475-0004